BEYOND THE

M**O**ON

A Conversational,
Common Sense Guide to
Understanding the Tides

BEYOND THE MOON

A Conversational, Common Sense Guide to Understanding the Tides

James Greig McCully
Duke University, USA

World Scientific

NEW JERSEY · LONDON · SINGAPORE · BEIJING · SHANGHAI · HONG KONG · TAIPEI · CHENNAI

Published by

World Scientific Publishing Co. Pte. Ltd.
5 Toh Tuck Link, Singapore 596224
USA office: 27 Warren Street, Suite 401-402, Hackensack, NJ 07601
UK office: 57 Shelton Street, Covent Garden, London WC2H 9HE

British Library Cataloguing-in-Publication Data
A catalogue record for this book is available from the British Library.

BEYOND THE MOON
A Conversational, Common Sense Guide to Understanding the Tides

Copyright © 2006 by World Scientific Publishing Co. Pte. Ltd.

All rights reserved. This book, or parts thereof, may not be reproduced in any form or by any means, electronic or mechanical, including photocopying, recording or any information storage and retrieval system now known or to be invented, without written permission from the Publisher.

For photocopying of material in this volume, please pay a copying fee through the Copyright Clearance Center, Inc., 222 Rosewood Drive, Danvers, MA 01923, USA. In this case permission to photocopy is not required from the publisher.

ISBN 981-256-643-0
ISBN 981-256-644-9 (pbk)

Printed in Singapore by Mainland Press

Dedication

This work is dedicated, in memoriam, to my mother, Phyllis Greig McCully. Although she probably would not understand most of the science in this book, she did understand something much more important, that the moon and the ocean are gifts for which we should have a reverent appreciation and gratitude.

" ... the ancients believed that the earth was on the back of an elephant, that stood on a tortoise, that swam in a bottomless sea. Of course, what held up the sea was another question. They did not know the answer.

The belief of the ancients was the result of imagination. It was a beautiful and poetic idea. Look at the way we understand it today.

This universe has been described by many, but it just goes on and on, with its edge as unknown as the bottom of the bottomless sea of the other idea — just as mysterious, just as awe-inspiring, and just as incomplete as the poetic pictures that come before."

Richard Feynman
Nobel laureate, physics.

Preface

"I am sorry that this letter is so long. I did not have time to write you a shorter one."

George Bernard Shaw

During the early years, Richard Feynman played bongo drums, pursued beautiful women, and helped build the first atomic bomb at Los Alamos. Later he taught at Stanford University, continued on the bongo drums, and won the Nobel Prize in physics. As a college professor, he became a great teacher in spite of his handicap (he was a genius). One of his students related asking another professor the question, "From a scientific and engineering viewpoint, what exactly *is* friction?" The answer was a barrage of mathematical equations. He complained to his roommate that he understood the math, but he still didn't exactly know how friction worked. He was advised to ask Dr. Feynman.

This time, instead of getting calculus, he was engaged in a conversation. Feynman began talking about a man slipping downhill on a sidewalk. Of course, the man's forward motion created the simplest form of kinetic energy. Then he described how the tiny grains of the concrete surface would pull off bits of shoe leather, slowing the motion of the man. He explained that each time a fragment of leather was pulled off, some energy was required, and this was subtracted from the man's kinetic energy downhill. "That's friction," Feynman concluded. His students were very fortunate, indeed.

After the Challenger space shuttle disaster, Feynman was a member of the commission to investigate the explosion on lift-off. In the first several days of hearings, engineers from the civilian contractors that built the rocket offered a cacophony of conflicting and confusing testimony. It did arise from this smoke screen that the lift-off had been delayed while NASA debated the risk of a launch during very cold weather. The first-stage rocket booster had been sealed with rubber gaskets which had a limited tolerance for cold. On the ride from his hotel to the hearings, Feynman had his taxi detour by a hardware store, where he purchased a strip of black rubber. He cut the rubber strip into two short lengths. Next to his microphone in the commission hearing-room, there was a pitcher of ice water. He dropped one piece of rubber into the pitcher before the hearings began.

Again, the highly technical and encoded testimony was unintelligible to the lay public. Finally, the chairman asked Dr. Feynman for his opinion. He said, "I think *this* is what happened." He lifted the length of room-temperature rubber, and flexed it back and forth. Then he reached his hand into the ice water, and held up the piece of cold rubber. He snapped it in two, and gravely affirmed, "*That's* what I think happened." Ultimately the commission concluded that the first-stage booster rocket exploded because the gaskets failed at low temperature.

Feynman wrote a number of books in the science-for-laymen genre. I have read most of them, and once came across a chapter on gravity. He addressed a question about the gravitation of the moon and semi-diurnal tides (two high tides and two low tides each day). The tides are of critical concern to me, as my passion is saltwater fly-fishing. Feynman's chapter included a stylized image of the moon, the earth, and its oceans. See illustration 1, below. Notice that there are two bulges in the earth's oceans. One bulge is displaced directly toward the moon's gravitation. The other bulge is on the opposite side of the earth, directly away from the moon.

Uncharacteristically, Feynman's explanation failed to satisfy. I was left with this irksome image of the ocean bulging up away from the moon for no apparent reason. This created one of those urgent and obsessive mental specters that haunt you — like a name you should remember, but cannot. I had no doubt that this fact was accurate because of the source. Furthermore, it was clear that this would explain two high tides per day; since the earth revolves around each twenty-four hours, any given point on the earth's surface would necessarily pass under the two bulges daily. However, I was left to obsess over the explanation, and as my wife will attest, I am fiercely obsessive.

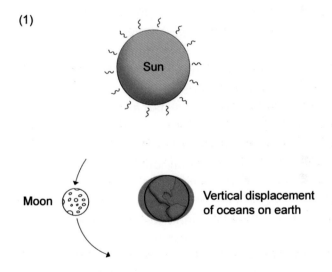

I began to ask professional saltwater fishing guides and yachtsmen why there was a displacement of the earth's oceans directly away from the moon which caused the semidiurnal tides. Some of them informed me that this simply was not possible, and obviously incorrect. I sided with Feynman, the Nobel laureate. Other experts were sure that the gravitation of the sun caused the mysterious second bulge in the waters. I knew that this was not the case, as I was also researching the subject in the library.

I continued to search for the answer by questioning all sorts of experienced mariners, sailors, offshore fishermen, and shrimp boat captains. People started to avoid me at my local marina. All that I learned was that the world is full of people who use the tide tables every day, but do not understand the forces of nature that generate the tides.

Finally, I got my answer at a social gathering, which took place in the evening on board a sailboat. The host had the clever notion to invite an astronomer who would describe the heavenly constellations for the guests, much as you might hire a magician to entertain at a party. As luck would have it, the sky was completely hidden by clouds, and the astronomer was left to his own devices. It is not everyday that you are in the company of an astronomer, so I took this opportunity to ask my question once more. This time the answer was, "That's simple," and he proceeded to draw the bulge of water away from the moon on a cocktail napkin, and easily provided a lucid and satisfying solution.

It *is* a fact that there is a vertical displacement of the earth's oceans in the direction directly opposite the moon, as shown in illustration 1. It *is* this second bulge in the waters that explains the enigma: one moon overhead every twenty-four hours, one high tide every twelve hours. Don't worry if the fiendish obsession is already upon you. By the end of chapter two, you will have the answer that will exorcise this demon.

After I was given the answer to this riddle, I soon began a study of the tides in earnest. The "one moon — two tides" problem was soon replaced by other annoying tide patterns. In my local waters on the coast of northeast Florida, the twice-daily high tides (semidiurnal tides) are equal height on some days, and very unequal height on other days. Again, I interrogated friends who are fishing guides and sailors. Again, I learned that the mariners who depend on tide tables misunderstand the flood that floats them over the shoals and brings them the fish. By now, I had gained access to a local college library and befriended a professor of physical oceanography. I could not find any thorough discussion of the tides on earth within any one sourcebook written for laymen. In order to expedite the process, I asked the oceanography professor for the title of such a book. To my surprise, he told me that there was no such comprehensive text for laymen. And so, naturally, I decided that the only thing to be done was to write one myself.

The answers started coming more easily. But, every time I thought that I was thoroughly knowledgeable, I would encounter another incomprehensible tide table. On a trip to the Gulf of Mexico, I was dismayed to learn that there is only one high tide each day in Panama City, Florida, only two hundred and fifty miles from Jacksonville, Florida, which has semidiurnal tides. To make matters even worse, the tables in Naples, Florida, inform boaters of one high tide in twenty-four hours on Monday, for example, and then two high tides on Friday of the same week! So, it was back to the library to solve two more pieces of the puzzle. All of this reminds me of playing golf — every time you feel competent the damnable thing makes you humble again (It also reminds me of the time we tried to bring down bats with a shotgun — but, that's another story).

Eventually however, the mysterious tables yielded their secrets. I found that the tides are not as chaotic as the weather. You can enter the barometric pressure, air temperature, humidity, and wind velocity into a computer, but you will still only have about a sixty percent chance of predicting the weather next weekend, three days away. By learning the forces of nature that create the tides, you can enter data into a computer and generate a reliable tide table for next year. The forces of nature that cause the tides are finite, logical, and accessible to analysis and comprehension by every mariner. Some tide patterns are complex, but they are not mysterious. The formulas for the tables are not confined to mythical Neptune's submarine cave. They are not even stamped

"Top Secret" in a laboratory at Wood's Hole. All tide tables make sense, and anyone who can set a sail or set a hook has the sense to understand them.

A word to the wise: read on patiently. Recall the last time that your vessel was on a sandbar. After the first hour, the tide was rising inch-by-inch, but you didn't feel any less aground (either you are afloat, or you are not afloat). You will be able to explain all tide tables after the last chapter, and not before. This is one of those complex subjects that require a gradual accumulation of facts, until suddenly you are no longer aground.

Science does not need to be obscure to laymen. Before we move on to our topic, the variety of the tides on earth, here is one last example of how it is possible to gain a profound understanding of nature without mathematics, and without the arcane abstraction of advanced science. Buckminster Fuller was one of the great problem solvers of the twentieth century. He was an innovator who found new solutions to old problems. One of his contributions to structural engineering was the geodesic dome. One evening he was sitting before his fireplace, enjoying the warmth of the burning logs with his five year old granddaughter. The little girl asked him, "Granddad, what *is* fire?" Without hesitation he replied, "Those logs came from trees. For many years the bright, warm, yellow sunshine was taken into the leaves of the tree, and the tree stored the sunshine in the wood. Now, we are letting that same bright, yellow warmth back out of the logs and into our house." Buckminster Fuller once said, "When I am looking for a solution to a problem, I never think of beauty. I only think of how to solve the problem. However, when I am finished, if the solution isn't beautiful, then I know that it is wrong."

It is time to invoke the spirit of Richard Feynman, or at least resolve to follow his example. Trust me, the libraries are full of books and chapters on the tides reduced to calculus. To really understand this subject, what is needed is something more like a conversation.

Acknowledgments

I am greatly indebted to Dr. Jim Beets for his encouragement in the early days of this project. Then, Dr. Kevin Bodge generously provided guidance and insight into the scientific pursuit of tide theory. Mrs. Julie Thordarson McGuire directed me toward a lucid and readable syntax. Dr. F. E. Wylie's book, *Tides and the Pull of the Moon*, excited me about this subject. Mr. Calvin Atwood convinced me that my enthusiasm would make up for my ignorance. Dr. Will Strunk's book, *Elements of Style,* reminded me to be concise, and counseled me to write from within myself, to write just the way I think. At a time when I needed a boost, Mr. Jonathan Eaton made me believe that this effort was worthwhile. Dr. John Klauder helped me through a rough patch, and reassured me that I was on the right track. Steven Gill, at NOAA, provided professional confirmation of material that I had pieced together as an amateur. Dr. Robert Cocanougher pretended that my hand drawn illustrations were not pitiful, and then led me to Ms. Joanna Muñoz, who artfully illustrated *Beyond the Moon*. Dr. Stanley Liu, a rare combination of intellectual brilliance and kindness to strangers, was first my mentor and then became my friend. The staff at World Scientific Publishing Company held me to rigorous standards of scientific accuracy, kept me humble, and improved this work. Mr. Bob Smith provided the perspective of a first time reader with a well-trained mind and a keen eye. My wife, Marion, was my respected critic, my faithful supporter, and on this voyage, as always, both my anchor and my compass.

Contents

Preface ix

Acknowledgments xv

1. The Tides in History. The Challenge of Understanding the Tides on Earth 1
2. One Moon–Two Tides 15
3. The Lunar Orbit 31
4. The Solar Influences and Solar–Lunar Interaction ... 45
5. Celestial Harmonics 59
6. The Coriolis Force and Oceanic Amphidromes. Coastal Kelvin Waves. Tidal Currents. Sea Level ... 71
7. The Seiche Effect and Basins of Oscillation. Tidal Intermixing 97
8. Coastal Geography and Near Shore Topography, Resonant Co-Oscillation, Sustained Forcing 113
9. Shallow Estuaries and Tidal Bores 145
10. Computation of the Tide-Tables and Chaos Theory 157
11. The Weather and the Tides. Atmospheric Tides ... 185
12. The Tides and Saltwater Fishing 197
13. The Constituents of the Tides on Earth. Synopsis of Tidal Influences 215
14. Epilogue 243

Practical Definitions 259

Bibliography 271

Index 281

Chapter One

The Tides in History. The Challenge of Understanding the Tides on Earth

" … the earth sweats to create the oceans, which accounts for their saltiness."   Aristotle

" … the body of the earth has its ocean, which also rises and falls every six hours with the breathing of the world."   Leonardo da Vinci

By the summer of 325 B.C, Alexander the Great had extended his march of conquest all the way to the Indian Ocean. When Alexander was a boy of thirteen, his father, King Philip, had persuaded Aristotle to come from Greece and tutor the young prince. Historians have described this merger as, "genius meeting genius." Alexander acquired scientific knowledge only known to a small circle of Greek and Macedonian scholars who were fortunate enough to study under Aristotle, the great apostle of Plato. King Philip understood that knowledge was power. He was grooming his successor to venture out and bring the spoils of war back to Macedonia.

Later, while prosecuting a military campaign, Alexander learned that Aristotle had published his lectures for wider distribution. Though occupied with managing 15,000 troops, he sent an angry message to his tutor: "In what shall I surpass other men, if those doctrines wherein I have been trained are to be all men's common property?"

A ruthless warrior and a brilliant strategist, Alexander's victories made him flagrantly egotistical. He promoted the myth that before his birth, his mother dreamed that she was, "penetrated by lightning, and her womb gushed fire, which spread around the world." Nonetheless, he always knew that Aristotle's lessons had provided him with a distinct advantage over his fallen adversaries.

Alexander launched his army aboard a fleet of thirty-oar galleys, in order to cross the broad delta of the Indus River. At a point near the open sea, a sudden monsoon forced them into the safe harbor of a side channel. This temporary setback would not have concerned the seasoned military man — he had traveled 17,000 miles and crossed many rivers, although he didn't know how to swim.

What *did* disturb Alexander was that all of the water gradually disappeared from the creek, leaving his immobilized army at the mercy of his enemies. His biographers report that the officers wandered about the mudflats, avoiding giant crabs and other unpleasant creatures.

As Mediterranean sailors with a meager knowledge of the tides, they had no reason to expect that the water would ever return. Their general had never seen an ocean before, and had no idea why the creek had gone dry in only a few hours.

Then the stream began to fill with water again. In this particular estuary, the incoming tide is an alarmingly rapid "tidal bore" (see Chapter 9). I suspect that Alexander had mixed emotions about this new flood, which first mobilized his army, and then sank his galleys damaged by grounding. As a general rule, tidal bores are not welcomed by men wearing uniforms of leather and iron. He must have been thinking, "Oh, this is terrific. Old Aristotle forgot to mention one small detail: that the entire ocean goes up and down ten feet, twice a day!"

The "cradle of civilization" was a mixture of Arabic and Greek science and philosophy. The Renaissance, stretching from Leonardo da Vinci to Galileo, began in Italy. Early western civilization was completely ignorant of the tides, because all of this took place on the shores of the Mediterranean Sea, which as it happens, has no significant tidal range.[a] This will be explained in chapter eight. The ebb and flow in this vast sea is only a few inches in twenty-four hours, barely noticeable among the waves on the shore. Thus, the early scientists and seamen did not misunderstand the tides — they simply never gave it any more thought than the residents of Tahiti gave to snowflakes.

Soon however, early navigators ventured out of the Mediterranean, to the British Isles and beyond. In an interesting twist of fate, civilization departed one of the least tidal ranges on earth (the Mediterranean — a few inches) and soon sailed into one of the greatest tidal ranges on earth (the Bristol Channel, England — over thirty feet). This unexpected phenomenon demanded an explanation. These early Greek and Italian explorers were quick to see a relationship between the tides and the phases of the moon. Of course they were unable to understand this relationship, since astronomy was still astrology, and everyone believed that the earth was the center of the universe. We can safely say "everyone," since all dissenters had been burned at the stake.

It would be fourteen centuries before Copernicus's grand theory of the solar system, and another two hundred years until Newton connected the orbits with gravity. For ten centuries before these accomplishments, the general consensus was that the moon passing near the earth compressed the atmosphere, which pushed down on the oceans and set them in motion, bobbing up and down.

---

[a] The range of the Mediterranean tides is less than 12 inches on most shores. There are pockets of higher tides in the Mediterranean, such as the three foot tides that occur at Venice, Italy.

During the middle ages, some of the greatest minds in Europe took the logical approach that since light was the only thing extending from the moon to the earth, the tides must be related to moonlight. It was widely accepted that: (1) the angle of the moonlight on the surface of the ocean varied at different times of the month, and (2) this light warmed the ocean, and (3) the resulting heat released gases trapped in the depths of the ocean, which (4) lifted the surface and created the tides.

Even the great Galileo, who understood that the moon orbited the earth orbiting the sun, misunderstood the tidal motion. He was struggling with the tide problem while on a ship passage in the Adriatic Sea. Foul weather set the boat decks pitching about, and Galileo noted that the ship's cargo of fresh water was agitated into waves within the water-vessels sitting on the deck. Observation: the motion of the ship's deck causes waves in the fluids onboard. Conclusion: The tides are due to the oceans being set in motion by the rotation of the earth. He reasoned that the earth's motion caused the oceans to move, much as a bowl of soup loses its equilibrium when the waiter sets the tray in motion across the room. He even considered that the tides were proof of the rotation of the earth. What was needed, of course, was gravity. And so, Isaac Newton was born in the same year that Galileo died.

Think back on what you were doing in your early twenties. Now consider this. In Newton's early twenties he concluded, by pure intuition, that the moon and the earth attracted each other in exactly the same way that a falling apple is attracted to the earth. In order to explain why the moon stayed up there, he developed his Laws of Motion. The prior generation (Galileo and Johannes Kepler) had provided him with a moon *in motion*, in orbit around the earth. Little Isaac learned this in school.

Every schoolchild knew that a ball thrown horizontally would not fall to the ground until it lost its motion. After school they would throw rocks with a sling, like David used to slay Goliath.

To the other boys this was a game. To young Newton this was an experiment, proving that although a rock will fall to the earth, a rock in circular motion in a sling will, "stay up there," so to speak.

By the time Newton was forty-five, he had derived the simple but profound mathematics describing the exquisite balance of gravitational attraction and the momentum of motion, which we all take for granted today. For those readers who don't really understand the true nature of gravity: how does it work from a distance? — it may be some consolation to know that Newton was also unable to fathom this mysterious force. The greatest scientific treatise of all time was his *Philosophiae Naturalis Principia Mathematica*, in which he finally concluded: "To us it is enough that gravity does really exist and acts according to the laws which we have explained, and abundantly serves to account for all the motion of the celestial bodies and of our sea."

Principia was published in 1687. We still do not know the true nature of gravity.[b] However, after *Principia,* everyone soon became comfortable with the idea that gravity causes the liquid oceans to bulge up (or out) toward the moon and the sun. Our intuition is easily satisfied that the clocklike cycles of celestial orbits, and the ponderous revolving of our planet on its polar axis, explain the rhythmic rise and fall of the tides on our shores.

The first systematic theory of the tides on earth was the Equilibrium Theory, first supported by Newton's new gravity. This was a simplistic understanding of how lunar and solar gravitation cause an elliptical deformity in the earth's oceans, with no consideration of the complications caused by the rotation of the earth nor the geography of the continents. Newton was followed by Lord Kelvin, who advanced the new science of hydraulics. Next, Bernoulli in 1740, and then Laplace in 1775, refined the mathematics and the physics of waves into a Dynamic Theory of

---

[b] See the supplement to chapter one, at the end of this chapter.

Tides. It accounted for the interaction of celestial forces and earthbound influences due to the interruption of tidal movement by the continents on a rotating planet, as well as the hydraulics over the continental shelf. Scientists began to envision the advancing tides as great oceanic waves, measuring 12,000 miles from crest to crest, moving westward around the globe.

In 1833, the British Admiralty published their first tide tables. These tables were largely empirical. That is, they were based on a series of measurements of recent tides around the calendar and around the world. They showed that the actual measured tides were manifest in many different patterns on diverse shores. It was abundantly clear that neither the pattern nor the height of the tide in any given bay or harbor could be predicted solely by the gravitational attraction of the moon and the sun. The early attempts at the Dynamic Theory were also insufficient to predict future tides. The ability to predict future tides by calculation would have to await another century of scientific progress. Eventually, twentieth century oceanographers condensed this learning into a cohesive and comprehensive model, which is the subject of this book.

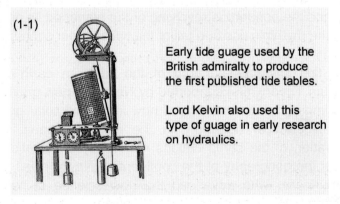

(1-1)

Early tide guage used by the British admiralty to produce the first published tide tables.

Lord Kelvin also used this type of guage in early research on hydraulics.

In the early twentieth century, a model called Response Analysis attempted to explain the complexity of the tides on the basis of input-output through a "black box." The celestial forces provided the input into the black box, which was the oceans; the output was the tide patterns. The current model of the tides on earth is the Harmonic Theory. The "pure harmonic model" was developed by Doodson and Proudman, and is the most commonly used model in recent decades. It explains the tides as the sum of a finite number of simultaneous, independent, sinusoidal (waveform) constituents, with the *frequencies* determined by the astronomical forces, and the *amplitudes* resulting from the effects of oceanography, shallow water hydraulics, and coastal geography. A revised set of mathematical formulas for tide prediction, known as the "almost harmonic" model, was offered by George Darwin. This sounds suspiciously like the theory that my wife uses when singing in church. It accounts for more variation in the astronomical forces, which repeat in a cycle every 18.6 years. Although it is theoretically an improvement over the pure harmonic model, it has not supplanted the more straightforward mathematics developed by Doodson. The Harmonic Theory is the model used by most oceanographers, the basis of tide prediction at NOAA, and the subject of the following chapters.

One of these oceanographers, Dr. F. E. Wylie, constructed a wonderful analogy between the tides and music. To state that the tides are caused by the gravitation of the moon[c] is only true in the same sense that the music of a symphony orchestra is caused by the breath of the musicians, and the movement of their arms and fingers. It would be more accurate to say that orchestral music begins with and takes its energy from the breath and action of the musicians.

---

[c]There is a subtle difference between gravitation and gravity. Gravitation refers to the force of attraction between two masses, the force described in Newton's Law: **G = mass$_1$ × mass$_2$ / distance^2**. Gravity is a more specific term for the force(s) that act on a mass within the sphere of influence of the earth, the force(s) that determine the *weight* of an object. For an object on earth, such as an apple on a tree, this involves the gravitation of the earth, the gravitation of the apple, and the centrifugal force arising from the earth's rotation (more precisely, the apple's rotation because the tree is attached to the rotating surface of the earth).

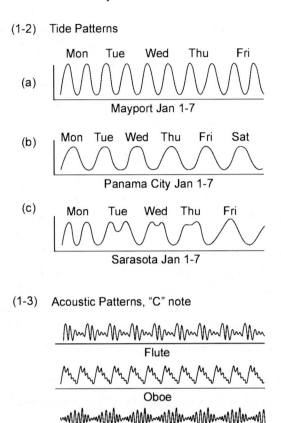

(1-2) Tide Patterns

(a) Mayport Jan 1-7

(b) Panama City Jan 1-7

(c) Sarasota Jan 1-7

(1-3) Acoustic Patterns, "C" note

Flute

Oboe

Clarinet

In an overly analytical sense, music is highly organized sound waves, passing through the air from the instruments to our ears. The *energy* of these sound waves *is* taken from the breath and motion of the musicians, just as gravitation sets the oceans in motion. But when the breath enters a flute, the music is very different from a trumpet. When the motion is exerted on a violin, the sound waves are very different from a cello. And, when the great energetic wave of the oceanic tide reaches the coastline of a continent, with its myriad beaches and bays, the tide is played out in myriad patterns.

The global tides derive all of their energy from the gravitation of the moon and sun and from the motion of the earth, while the complex and various patterns of the tides on the shores of different bays, gulfs, and seas are analogous to the various patterns of sound from different musical instruments, all set in motion by the same breath, the same energy.

This is a useful metaphor, to be sure. Illustration 1-2 shows the tide tables for several points on the coast of Florida, all on the same day, and all resulting from the same vertical displacement of the Atlantic Ocean moving westward. Illustration 1-3 depicts the sound waves of three different wind instruments playing the "C" note. These different patterns are all energized by moving air, which is essentially the same as it *enters* an oboe or a clarinet. However, the acoustic waves *within* each instrument are unique, because the instruments are different lengths and shapes, and made of different materials, just as the Mediterranean Sea and the Bay of Fundy have unique geography and hydraulics.

Hold onto this image, and carry it with you as we proceed together to investigate the forces of nature that influence the tides. I understand perfectly well that what you really want is the "how and why." How can the Bay of Fundy have a tidal range of fifty feet on the same day that Cape Hatteras's tides are two feet, all on the same east coast of the same continent? Why is there one high tide every twenty-four hours in the Gulf of Mexico, and two high tides each day on the opposite side of Florida, only 250 miles away? See illustrations 1-2(a) and (b). How can there be one high tide in twenty-four hours on Monday, and two high tides on Friday, at the same point on the coastline? See illustration 1-2(c).

The following chapters will decipher all of the curious and confounding tide tables that you will ever encounter. Never again will you endure this conversation on your vessel. First Mate, "Charlie dear, why are the tides so much higher here than at home?" Captain Clueless, "The tides are caused by the moon, sweetheart." Mate, "But they have the same moon here, Charles." Captain C., "Not exactly dear. The moon ... and, oh yes, the sun ... the sun is also involved ... and ... they have a different angle in the sky here." Mate: "Are you sure about that, captain?"

Captain Clueless might be facing a mutiny if his first mate knew this one fact from the following chapters. In the twentieth century we know the exact mass of the sun, moon, earth, and ocean. We have measured the exact distances between them and the strength of their gravitational fields. We now know that **the gravitation of the moon is only capable of displacing the water on the surface of the earth by approximately twelve vertical inches. The gravitation of the sun can only displace the oceans by approximately six vertical inches. Therefore, when the moon and sun are aligned to pull in the same direction over the center of the Atlantic Ocean, the maximal tidal range (due to the moon and sun) is approximately eighteen inches. The corollary to this fact is that all tidal ranges greater than eighteen inches must be due to something other than the gravitation of the moon and sun.**[d]

Before we begin a detailed study of the natural forces that explain all tide tables, let's fast-forward from the British Admiralty to the National Oceanographic and Atmospheric Administration for one more glimpse of what's in store. When the decision was made to drill for oil in Alaska, the United States government wisely decided to prepare very precise tide tables for Prince William Sound and its tributaries, where the tankers would remove the hazardous cargo. The computers were programmed with 114

---

[d] The scientific fact that the lunar and solar gravitation can cause only 18 inches of vertical displacement of the earth's oceans is true for a stationary planet (not rotating) which has a featureless surface (no continents). This is what the Equilibrium Theory predicted, but it is only true in the center of the deep oceans, and it does not hold up at the continental coasts.

independent variables which have influence on the tides — that is, the gravitation of the moon and sun, and 112 other important constituents necessary for these precise tide tables.

In the calculation of the standard tide tables issued annually to the boating public, The National Oceanographic and Atmospheric Administration includes 37 major independent constituents which have a measurable influence on the earth's tides. Again, the gravitation of the moon and sun, and 35 other factors. Before you conclude that the Prince William Sound tables were unnecessarily meticulous, you should know that by including 114 variables they ignored 282 other known influences on the earth's tides. The total is 396, according to the prominent oceanographer, Dr. Arthur Doodson.[e] These other factors are, of course, the "how and why" the patterns of the tides are as variable as the instruments in a symphony orchestra. Likewise, these factors are the "something other" than gravitation that causes all tides greater than 18 inches.

This book does not require 396 chapters. You can make sense of every tide table you will ever encounter using high-school science and without mathematical equations. Incidentally, you can stay off of the rocks or plan a fishing day without understanding the tables. By the same token, you can cross the Pacific Ocean even if no one aboard knows how to repair a diesel engine. But it is better to know; and there is an inherent satisfaction and pleasure that comes from mastering something as primal as the tides. I don't know why. It's just human nature.

---

[e] Dr. Arthur Doodson dedicated his life to the practical application of science in the prediction of the tides. He was the oceanographer chosen to predict the tides for the D-day invasion during World War II.

Supplement to chapter one:

You may be thinking, "Didn't Einstein figure out gravitation?" Well, Einstein said that gravitation was a distortion of the space–time continuum, where straight lines were curved lines, and the difference between velocity and acceleration was relative to the observer. Sorry you asked?

Actually, it is possible for us to gain some understanding of what Einstein was talking about. He wanted his revolutionary theories to be accessible to laymen, and he wrote about them in plain English. For example: A man is inside an elevator, which is accelerating upward at 12 feet/sec./sec. Another man shoots a bullet parallel to the ground, and the bullet passes through the elevator. The bullet enters the east wall of the elevator six feet above the floor. Because the elevator is accelerating upward, the bullet exits the west wall of the elevator two feet above the floor.

The question is: "What is the path of the bullet through the elevator — straight or curved?" The answer is that this depends on whether the observer is inside the elevator or outside. In other words, the path of the bullet can only be measured relative to an observer (relativity).

An observer outside the elevator would say that the bullet traveled in a straight line with a constant velocity, while the elevator accelerated through its path. A scientific observer inside the elevator would observe the bullet traveling in a curved path, accelerating toward the floor at 12 feet/sec./sec.

Is the bullet accelerating toward the elevator floor, or not? The answers is absolutely yes. Is this because the elevator floor pulls on the bullet from a distance? Absolutely not, but that is the only rational conclusion for an observer inside the elevator.

The point of Einstein's "thought experiment" (his term for this exercise) is that perhaps objects appear to be attracted toward each other according to the Law of Gravitation, because we observe these interactions within *our* framework of three dimensional space plus a dimension of time; but, there are other possibilities.

The idea that gravitation and magnetism can cause objects to act on other objects from a distance is a very peculiar idea. Einstein is saying that perhaps objects don't act on other objects from a distance at all; perhaps the nature of space and time make it appear to happen, just as the acceleration of the elevator creates the frame-of-reference of the observers inside. In other words, perhaps the nature of time and space constrains all objects with mass to interact with each other in a way that appears to be due to "action from a distance." This sort of explanation does not really provide an understanding of gravitation, but it does provide some insight into the nature of such a cosmic problem.

Chapter Two

One Moon–Two Tides

"Nothing is so certain to perpetuate ignorance
as failure to investigate the facts."   Ralph Waldo Emerson

The most common pattern of tides around the earth is semi-diurnal (two high tides, and two low tides, daily). As we shall see, there are basins around the globe with diurnal tides (one high tide, and one low tide, daily). On the coasts of many continents there is a "mixed" tide pattern, which is usually a mixture of the two, or even semidiurnal on some days and diurnal on others. Where the tides are semidiurnal, the two high tides are equal height on some days, and very unequal height on other days. Before we examine these complexities, we will deal with the common pattern of two equal high tides each day, consisting of about 6 hours of flood tide and 6 hours of ebb tide, twice daily.[a]

The tides are semidiurnal where I live on the coast of northeast Florida. They were semidiurnal when I was a boy, watching alligators in the St. Johns River. They are still semidiurnal when I fish in the salt marsh creeks near Amelia Island, Florida. For fifty years it never occurred to me that the two high tides each day required an explanation. When we have a novel experience or encounter something seldom seen, it sparks our curiosity. We crave an explanation for the rare and unexpected. But we take our daily experience for granted, as if there was no other possibility. Preschool children know that it is dark every night, but

---

[a] The period of time between successive lunar semidiurnal tides is 12 hours and 26 minutes. Therefore, the (average) duration of each flood and each ebb is 6 hours and 13 minutes.

they don't wonder why this is so. They never think about it at all — it's just the way it is, and it never occurs to them to question why. Adults in primitive cultures have theories for unusual phenomena such as an eclipse of the moon. However, they do not have theories for the commonplace events, such as breathing, gravity, or day-and-night.

When we begin our education, adults give us explanations for all sorts of things that we never thought required an explanation. We learn that the earth revolves on its polar axis every twenty-four hours, and this causes it to be dark when we are on the side opposite the sun. They teach us that the atmosphere absorbs certain wavelengths of light, and this causes the sky to be blue. I suppose that if we never encountered these ideas in our education or our reading, we would remain in childlike ignorance about the commonplace throughout our lives, like primordial adults.

When you consider the accomplishment of Isaac Newton in understanding gravity, his real genius was that he thought about it at all, that he thought it required an explanation. No other scientist was searching for a theory of gravitation. Everyone else in his time merely took it for granted. In this same way, I always took the semidiurnal tides for granted; and during my excessive (twenty-four years of) education, the subject was never mentioned. Of course, I learned in school that the tides are caused by the gravitational attraction of the moon. But I am quite certain that no one ever explained the semidiurnal pattern of the tides. And so, I set out to sea as ignorant as a child or a caveman.

When people say, "Ignorance is bliss," they mean the ignorance that is oblivious to the problem. There is another kind of ignorance. Once you become *aware* that you are ignorant, it is anything but blissful. Several years ago, I began to think about the tide-table for my local waters. See illustration 2-1. I made a simple diagram in order to be sure that I really understood why there are two high tides every day. See illustration 2-2.

(2-1)

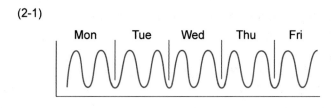

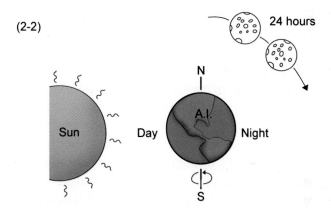

This diagram was my attempt to recreate the high school science version of the near solar system: the earth, moon, and sun. As I compared the diagram with the tide table, I became vaguely uncomfortable. Something wasn't right. I was becoming aware of my ignorance.

Looking at the diagram, I started with the facts that were certain: (1) The earth revolves completely around on its polar axis once every twenty-four hours. (2) The sun is overhead in a stable position during the twenty-four hours. Any given point on the earth's surface, such as Amelia Island, Florida, will rotate toward the sun once every twenty-four hours (we call this daytime) and rotate away from the sun once every twenty-four hours (nighttime). So far, so good. No problem here. (3) The moon also remains overhead in a fairly stable position, only moving about 1/28th of its monthly orbit during twenty-four hours. (4) Amelia Island rotates around toward the moon only once every day, just as it rotates toward the sun once a day. (5) The gravitation of the moon attracts the ocean water near Amelia Island, and causes the high tides.

This was the exact moment when I lost my childlike ignorance of the tides and I became just plain ignorant. Now, something was seriously wrong. My local waters revolve around toward the moon once every twenty-four hours; and yet, there is a high tide every twelve hours. I continued to study the diagram. Where is Amelia Island twelve hours after the first high tide? Obviously, it is halfway around its daily revolution, on the side of the earth directly *away from the moon.* This was not good for my mental equilibrium. Why would there be a high tide on the side of the earth facing away from the gravitation of the moon? There are many things that I don't understand: what happened before the "big bang," how do gravitation and magnetism act on things from a distance, why do so many people watch daytime television? I don't worry about every unanswered question. But, this ridiculous situation — all of a sudden, I didn't even understand why there were two high tides every day on my local waters — this was intolerable. Some of my friends are professional saltwater fly-fishing guides, who depend on the tide to make a living. I expected them to explain away the enigma. I got answers, but they were unconvincing, and everyone had a different solution.

It is a truism that if you chase something too hard, it will elude you; but if you stay quiet, it may come to you. And, sure enough, a few months later I was reading a book on another subject altogether, and part of the answer came to me. In the introduction, I mentioned that one of my favorite authors is a

Nobel Prize laureate, the American physicist, Dr. Richard Feynman. In a chapter on gravity, he included an illustration which caught my attention. See illustration 2-3.

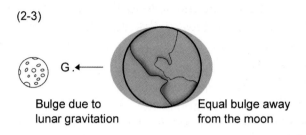

(2-3)

Bulge due to lunar gravitation

Equal bulge away from the moon

Feynman explained that there are always two bulges in the earth's oceans. One of these is always toward the gravitational attraction of the moon. The other is a vertical displacement of the water, which is always directly away from the moon. It required some further reading and the interrogation of an astronomer to understand the cause of this fact, which explains the two high tides every day.

In order to thoroughly understand this phenomenon, we will approach it from the viewpoint of the astronomer who dispelled my ignorance. If you look into space, you will find many objects in orbit around other objects. There are three possible orbital systems. In illustration 2-4, there is a miniscule object, like a grain of sand, orbiting a massive planet like Jupiter.

Because the gravitation of the tiny moon is negligible, the planet rests motionless in the center of the moon's orbit — that is, in the axis of the orbit. This is the simplest form of orbital mechanics. This is the simple-minded concept of the interaction between our earth and moon that we learned in high school. This concept causes our misunderstanding of the tides, because **the earth does not rest motionless at the axis of our moon's orbit.**

(2-4)
Tiny moon, huge planet.
Moon orbits planet.

The second possible orbital system is shown in illustration 2-5. In the universe there are systems with two bodies of equal mass in orbit around a common axis in space.[b] One example is binary stars. In these systems, two opposing forces are at work. Both objects are pulled toward the center by gravitation, and both objects are pushed out from the center by centrifugal force, according to the law that objects in motion tend to remain in motion (and continue in the same direction). As their centrifugal force constantly urges them to fly off in a straight path, their mutual gravitation constantly turns them back toward the center. Since they are equal in mass, and the momentum of their motion is exactly equal to their gravitation, they will remain in a stable orbit around an axis point in space.

(2-5)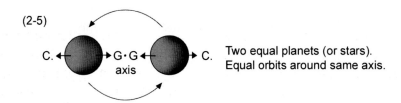
Two equal planets (or stars).
Equal orbits around same axis.

---

[b] When two objects are orbiting the same axis, the axis is the center of their combined mass. If the two objects are exactly the same mass, this center (the axis) is exactly equidistant from both of them. If they are of different mass, this center is closer to the more massive one.

The third possible orbital system is shown in illustration 2-6. In this example, a planet like our earth is orbited by a moon that is massive enough to have its own significant gravitation. The gravitation on the surface of our moon is 1/6th of the earth's surface gravitation. At this point, it is important to remember that the Law of Gravitation does not state that large objects attract smaller objects. The Law of Gravitation expresses the fact that all objects attract all other objects.[c] Therefore, the earth and the moon attract each other. If the earth was at rest at the axis of the moon's orbit, this mutual attraction would eventually pull the two bodies crashing into each other.

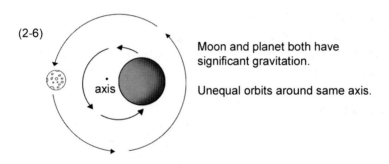

(2-6)

Moon and planet both have significant gravitation.

Unequal orbits around same axis.

In order for the orbital system to be stable and sustainable over eons of time, some other force must balance this attraction. The moon maintains an orbit around an axis point in space, because the moon's centrifugal force outward balances the gravitational attraction toward the earth. But, again, the earth would be drawn toward its massive moon unless some other force acted on the earth and balanced the gravitational attraction of the moon. What is this other force? Return to the illustrations. If the mass of the moon is insignificant, as in illustration 2-4, the planet can, for all practical purposes, remain at rest. If the two bodies have the

---

[c] **Force of gravitation = mass$_1$ × mass$_2$ / distance^2**.

same mass, as in illustration 2-5, they both must be in motion, in the same orbit, around the same axis point in space, so that centrifugal force can act on both and balance their mutual gravitation.

Illustration 2-6 depicts an orbital system that is somewhere between these two extremes. This is the orbital system of our earth and our moon. And so, **our moon orbits an axis-point in space and remains in stable orbit because of its centrifugal force; and the earth also orbits the same axis-point in space and remains in stable orbit because of *its* centrifugal force.** The objects in illustration 2-5 are of equal mass and have equal orbits. The objects in illustration 2-6 have unequal mass, like our earth and moon, and therefore have unequal orbits.

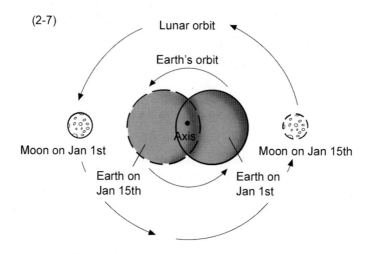

Radius of lunar orbit — 250,000 miles
Radius of earth orbit (of the lunar orbital axis) — 3,000 miles
Diameter of earth — 8,000 miles
Circumference of earth orbit — over 20,000 miles

In fact, the radius of our moon's orbit is about 250,000 miles, and the radius of the earth's orbit around the same axis is about

3,000 miles. See illustration 2-7. Since the diameter of the earth is approximately 8,000 miles, the axis of its orbit is inside the earth, about 1000 miles below the surface. This does not imply that the earth's motion is negligible, however. The circumference of the earth's orbit around the lunar orbital axis is over 20,000 miles, and it travels this orbital path every 27.5 days. That much motion by an object as massive as our planet creates a lot of centrifugal force: exactly enough to balance the gravitational attraction toward the moon.

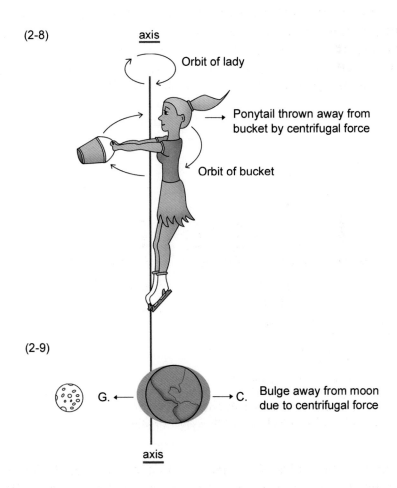

Illustration 2-8 depicts a lady spinning around on ice skates. She is holding a bucket of water at arms length. The weight of the bucket, her "moon," is supported by centrifugal force as it "orbits" her. Notice that her skates are on an axis line drawn through the center of this orbit. Her upper body, however, must lean back away from the bucket, in order for her to balance its weight, and avoid falling forward. This causes her body to move in its own smaller orbit around the axis line. Finally, notice that *her ponytail is thrown out by her centrifugal force, in a direction directly away from the bucket.*

Now examine illustration 2-9. It should be clear that **the bulge of water on one side of the earth is caused by the gravitational attraction of the moon, while the bulge of water on the other side of the earth is caused by the centrifugal force of the earth's orbit around the lunar orbital axis.**

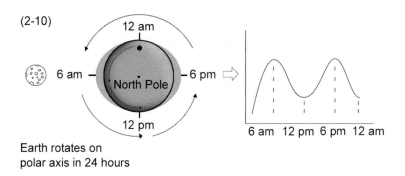

Earth rotates on polar axis in 24 hours

To complete the picture, we only need to add the daily rotation of the earth on its polar axis. Illustration 2-10 looks down at the earth's north pole. As any given point on earth revolves around on its latitude once every day, it first passes through the bulge caused by lunar gravitation, and then, twelve hours later, it passes through the bulge[d] caused by centrifugal force. And voila! Two high tides every twenty-four hours.

It is important to distinguish between the mechanism above and the separate and distinct centrifugal force related to the earth's daily rotation on its polar axis. The centrifugal force associated with the earth's rotation on its polar axis[e] is equal at all points around the globe and plays no role in the formation of the bulges that form the tidal waves, the "tidal ellipse" of the earth's oceans.

It is commonplace to think of the tidal high water as a wave moving around the globe, approaching our coastline from the east toward the west. This is an accurate description, but it is also accurate to say that the displacement of the water remains in fixed alignment with the moon, while the earth revolves our coastline from the west toward the east. We can show that both statements are equally valid. To an observer standing on the earth, the wave of high water approaches him from east to west, as he stands still on the coastline. To an observer on the moon, the high water always remains aligned toward him on the moon, while the coastline revolves from west to east. As we expand our investigation into the complexities and variations of the tides, we will alternate between these two images, depending on the context, in order to shed the most light on the subject of that chapter.

Now that you have a more accurate description of the orbits of the earth and its moon, you have all of the information needed to understand why there are two high tides every twenty-four hours: one aligned toward the moon due to lunar gravitation, and one

---

[d] Oceanographers do not use the term "bulge" to describe the deformity of the earth's oceans by lunar and solar gravitation. The correct scientific terminology is, "lunar tide ellipse" and "solar tide ellipse." This indicates that the oceans are deformed from a sphere to an ellipse by gravitation and centrifugal force.

[e] See the supplement to chapter two, at the end of this chapter.

aligned away from the moon due to the centrifugal force of the earth's orbit around the lunar orbital axis. You can skip ahead to the next chapter, and ignore the following detail without sacrificing anything vital to your mastery of the tides. However, I suspect that if I omit this one final detail, a certain minority of readers, perhaps trained in mathematics or engineering, will detect a flaw in the above discussion and lose faith in this entire exposition. And so, having stated that the following is, in a sense, superfluous, I will include this detail for the perfectionists in the crowd.

A stable orbital system depends on a precise balance of gravitational and centrifugal forces. One measure of this symmetry is that the force causing the vertical displacement of water toward the moon is exactly equal to the force causing the vertical displacement away from the moon. You might ask the question, "Since the earth is about eight thousand miles diameter, the gravitation of the moon should be weaker on the *far* side of the earth than on the *near* side. Why doesn't this cause an asymmetry?" This is a good question.

Illustration 2-11 depicts several principles essential to the answer. To begin with, the force of gravitation and the centrifugal force are only equal and balanced at the exact center of the earth. They are not equal on the near side of the earth, nor on the far side of the earth. This is because the force of lunar gravitation is inversely proportional to the square of the distance from the moon: when an object is twice as far away, the gravitation is only 1/4th as strong.

The side of the earth opposite the moon *is* farther away, and the moon's gravitation *is* weaker than it is on the side toward the moon. This is not the case with the centrifugal force involved in this particular motion of the earth (the particular motion shown in illustration 2-7), so that the centrifugal force is the same on both sides of the earth.

One Moon–Two Tides

(2-11) Lunar gravitation greater on side toward the moon. Centrifugal force equal at all points.

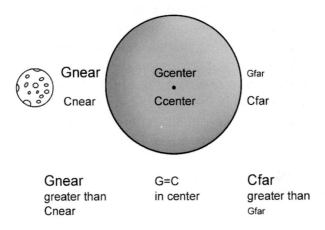

Using the annotations on illustration 2-11, we can summarize this:

G_{near} is greater than G_{center} is greater than G_{far}.
C_{near} equals C_{center} equals C_{far}.
G_{center} equals C_{center}.

Therefore,

G_{near} is greater than C_{near}, and
C_{far} is greater than G_{far}.

Although we can simplify the explanation for the semidiurnal tide pattern — one high tide due to gravitation, and one high tide due to centrifugal force — it is more precise to state that there are *net forces* on the two sides of the earth, which are aligned toward the moon and away from the moon, respectively. The net force on the side toward the moon is G_{near} minus C_{near} = net gravitational force$_{near}$. The net force on the side away from the moon is C_{far} minus G_{far} = net centrifugal force$_{far}$. In other words, gravitation dominates the net force on the side toward the moon, and centrifugal force dominates the net force on the side away from the moon. The truly remarkable thing about this "detail" is that **this difference between the lunar gravitation on opposite sides of the earth is the only reason for the tidal influence of the moon.**[f] This additional information does not in any way change the way we understand the semidiurnal tide pattern. It is just more precise.

---

[f] It is somewhat counterintuitive that lunar gravitation at 250,000 miles would be much different from lunar gravitation at 258,000 miles, and sufficient to be the sole cause of the lunar tidal influence. This is the case however. It may be easier to accept in concert with the following: the *net* gravitational force is relative to the inverse *cube* of the lunar distance, not the inverse square. This derives from the equation:

$$\text{Force}_{net} = \text{lunar } G \left\{ (1/D)^2 - 1/(D + r)^2 \right\}.$$

Therefore: $\text{Force}_{net} = 2 \text{ lunar } G \times r / D^3$

where **D** = distance between centers of earth and moon.
   r = radius of earth.

Again, it is not necessary to know this mathematical detail in order to understand that one tidal bulge is due to lunar gravitation, and one is due to the centrifugal force of the earth's orbit around the lunar orbital axis.

Supplement to chapter two:

The centrifugal force of the earth's rotation on its polar axis has no effect on the tide, as it is directed outward equally at all points on the surface. On the other hand, this centrifugal force affects the very shape of the earth. The earth is not exactly spherical. It is a deformed sphere, which has a greater radius from the center to the surface-at-the-equator, and a smaller radius from the center to the surface-at-the-poles. This shape is called an oblate spheroid. This deformity is caused by the centrifugal force of the revolving earth, which is greatest at the equator. It is possible because the earth is not rigid, but is actually somewhat elastic due to its liquid core and surface of tectonic plates. Also remember that the primordial earth was less solid and was also revolving faster, so that it would have been deformed by centrifugal force long ago.

A friend who shares my interest in scientific minutiae posed the following puzzler. If there is greater centrifugal force at the equator, wouldn't a person weigh less at the equator than at the poles? The answer is that *if* the earth was a sphere, the greater centrifugal force at the equator *would* cause you to weigh significantly *less*: about 10 pounds less for a 150 pound man. However, the earth is thicker at the equator and the greater gravitational attraction at the equator makes you weigh almost exactly the same amount *more*. Because the earth is not completely elastic, the centrifugal force at the equator cannot deform the globe quite enough to perfectly balance these forces, and, in fact, you will weight a tiny bit less at the equator: about 1/4 pound less for a 150 pound man. This bit of scientific minutia is just the thing to impress your buddies during the halftime break of a football game, or perhaps clinch your standing with a potential mate.

Chapter Three

The Lunar Orbit

"Thinking is more interesting than knowing,
but less interesting than looking."   Goethe

    I once spent some time with a big game hunter who had just returned from safari in Africa. He had been guided deep into the jungle to the remote village of an isolated tribe. The chief of this tribe, Nlovu, was extremely intelligent, although he was almost entirely unaware of the outside world.

    One evening, the twentieth century hunters and their primitive-culture guides sat around the campfire, bonding together. A full moon held sway in the night sky. The African natives asked if they could handle the hunting rifle my acquaintance had brought into their village. They admired this object as if it had arrived from another planet, or from the future. To their great delight, the natives were allowed to fire the rifle into the sky. The modern hunters had a few drinks, and then they did something they were not proud of later. They told the tribesmen that if they took careful aim at the very edge of the moon, they could knock off little pieces and see them fly off of the moon into the night sky. These intelligent but uninformed bushmen had no reason to doubt this contention. Taking turns, they tried repeatedly to knock off a fragment of the moon. When this failed, the hunters explained to them that their aim was faulty, and they were missing the edge.

    This took place after men from another continent had already landed on the moon. This anecdote illustrates how the great scientific thinkers like Gallileo, Newton, and Einstein have done much more than discover some new facts; they have completely redefined our place in the world.

(3-1)

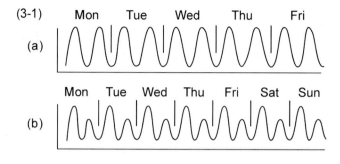

(3-2)

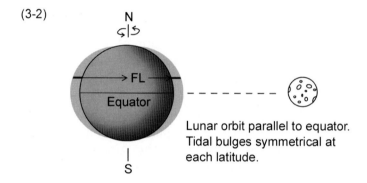

Lunar orbit parallel to equator. Tidal bulges symmetrical at each latitude.

Now that we understand the forces that cause the semidiurnal tide, we should ask why the two daily high tides are equal on some days, and very unequal on other days. See illustration 3-1. The table, 3-1(a), is for Mayport, Florida, for the week of June 1st through 7th. During this week, the two high tides on each day are *equal* height. The table, 3-1(b), is for Mayport, Florida, for the week of June 15th through June 22nd. Now, at a different time on the same shore, the two daily high tides are *very unequal* height. There is a simple explanation for this common phenomenon.

Illustration 3-2 depicts the moon in orbit around the earth, in a path that is parallel to the equator, or in the plane of the equator. Notice that the two bulges in the earth's oceans are greatest at the

The Lunar Orbit

equator, and they are symmetrical over each latitude as they proceed toward the poles. Therefore, the height of water is the same at each latitude, on both sides of the earth.

Now picture the earth revolving on its polar axis. Follow the latitude of northeast Florida (latitude$_{fla}$) as it revolves through the gravitational bulge, and then through the centrifugal bulge in the oceans. Because the tidal bulges are symmetrical to the equator, the water level at latitude$_{fla}$ is the same on both sides of the earth, and the two high tides in northeast Florida will be equal. This explains tide-table 3-1(a), but what about table 3-1(b)?

(3-3)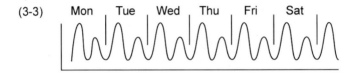

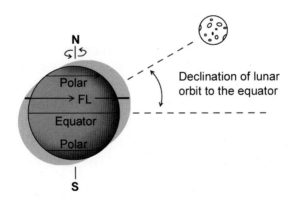

The orbit of our moon is not always in the same plane as the equator. The orbit may be in the plane of the equator, or it may be at an angle to the equator. See illustration 3-3. This angle is called the "declination of the lunar orbit." There is a rhythmic pattern of changing lunar declination between 28.5 degrees north of the equator to 28.5 degrees south of the equator, and like all of the astronomical influences on the tides, the lunar declination progresses in a repetitive cycle.[a] Illustration 3-3 shows how the moon's declination to the equator affects the semidiurnal tides. Again, follow latitude$_{fla}$ as it revolves for twenty-four hours through the two tidal bulges. Notice that the vertical displacement of the water is aligned with the moons orbit, and declined 28.5 degrees to the equator. Under these conditions, latitude$_{fla}$ will revolve through a higher high tide, and then twelve hours later, it will revolve through a lower high tide. This correlates perfectly with a semidiurnal pattern of unequal tides shown in illustration 3-1(b). Thus, **a pattern of unequal semidiurnal tides is due to the declination of the lunar orbit.**

There is another possible consequence of the lunar orbital declination. Look back at illustration 3-3, and you will see that the earth is annotated with north and south latitudes, "polar." Notice that only one of the tidal bulges reaches these far northerly and southerly latitudes when there is significant lunar declination. Therefore, the latitudes near the north and south poles will have only one high tide each day. Whereas the declination effect causes one high tide to be smaller near the tropic latitudes, it causes one high tide to be *absent* near the poles.

Looking at illustration 3-2, you might conclude that the tide should always be highest at the equator, and progressively lower as you proceed toward the poles. The lunar orbit is always near

---

[a]The declination of the moon to the equator changes in two distinct cycles: (1) Each month the moon passes over the equator twice, as the declination progresses from maximum northern to maximum southern, and then back to maximum northern declination monthly. (2) There is also a cycle which determines the extent of the maximal northern and southern declination. Every 18.6 years the maximal lunar declination cycles from 28.5 degrees to 18 degrees, and back to 28.5 degrees. This is called the "regression of the moons nodes." Its effect on the tide pattern is referred to as the "lunar nodal tide."

the equator, and the height of the vertical bulge *does* decrease progressively toward the poles. However, this makes a small contribution to the ultimate measured tide at each latitude. This vertical bulge in the oceans is only one of many factors that determine the tidal range at any location, and we do *not* find a direct correlation between height of the tide and the distance from the equator. By analogy, we do not expect to predict the outcome of a football game by choosing the team with the heaviest linemen — there are too many other players on the field.

Lunar gravitation is certainly the most important force that generates tidal energy. The lunar orbit has more influence on the tides than any other celestial cycle. However, the great variation in the height of the tide, measured on diverse shores, has little to do with the moon. The Canadians at the Bay of Fundy (who experience enormous tides) live under the same moon as the Italians (who experience no significant tides). Earthbound, non-celestial forces determine the variety of tidal patterns around the globe: the earthbound forces of oceanography, the hydraulics over the continental shelf, and geography are the likely source of the unique tide pattern in your homeport.

It is a commonly held belief among mariners and those who frequently use the tide tables that the range of the tides is higher as you progress toward the north from the equator. This is probably due to the 50 foot tides in the Bay of Fundy, Nova Scotia, the 35 foot tides in Ungava Bay, northern Ontario, and the 30 foot tides in Cooks Inlet, Alaska. This is not the case, however. Ocean chart 3-4, below, indicates the highest tidal ranges on earth as dark areas along the coastal zones. Notice that they are scattered randomly around the globe, with no particular north-south distribution.

During my research on the tides, I came across many interesting and surprising facts. The most surprising was that

the combined gravitation of the moon and sun is only capable of causing approximately eighteen inches of vertical displacement in the water on the surface of the earth. This scientific fact can be calculated from the appropriate physical laws, using the known mass of the sun, moon, and ocean, the strength of their gravitational fields, and the known distances between them. It has also been confirmed by measuring the actual height and range of the tide in the middle of the Atlantic Ocean. The tidal range can be measured in the middle of an ocean by earth satellite. This has also been measured by pressure sensors on the ocean floor, which can detect the amount of water above them. This mid-ocean tidal range of eighteen inches occurs over a six-hour interval, an average of three inches per hour.

(3-4)

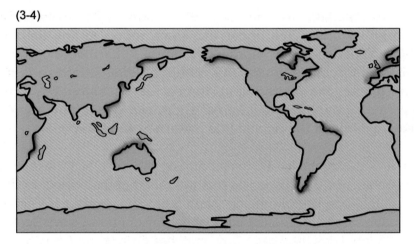

The dark shaded areas of the oceans are the highest tidal ranges on earth. Their distribution has no relation to the north-south latitudes.

When we are first confronted with this fact, it seems inconsistent with our personal experience. We encounter the tides on the coast of the continents, where we sometimes find alarming floodwaters and dangerous rip currents. On the shore of northeast Florida, where I live, the tide rises six feet in six hours.

The Lunar Orbit

We also experience precipitous falling tides, which threaten to strand our vessels, and damage our hull (and our pride). How can the paltry mid-oceanic tide cause such a dramatic deluge on our shores?

When a mass is displaced to a higher vertical position away from the earth, it acquires potential energy. You will recall that kinetic energy is due to a mass in motion (a locomotive going sixty miles per hour), and potential energy is due to gravitational attraction of a mass elevated to a higher point within the earth's gravitational field (a piano balanced on the roof of a dormitory). Water is very heavy. One cubic foot of seawater weighs 64 pounds. The displacement of the Atlantic Ocean, 18 inches higher, creates a lot of potential energy.

You can also compare the oceanic tidal energy to a mass falling down toward the earth, rather than a mass pulled up toward the moon. At the surface of the oceans, the gravitational field of the earth is 11,700,000 times stronger than the net lunar gravitation, since the earth is more massive and vastly closer to the surface of the oceans. It is this huge amount of water "falling" toward this powerful force that constitutes the potential energy of the tide. Now, consider that this enormous mass of elevated water is actually a wave racing around the globe. At the equator, the earth's surface is traveling at about 1050 miles per hour. The oceans are too shallow (13,000 feet, Atlantic, and 15,000 feet, Pacific) to allow a free moving deep water wave that is 12,000 miles from crest to crest. Therefore, the tidal energy is "only" traveling around the globe at about 430 miles per hour.[b]

---

[b] The speed of a deep water wave is governed by the depth of the basin, according to the formula: **v = the square root of (g times d)**, where **v** is the speed, **g** is gravitational acceleration, and **d** is the depth. Because the tidal waves have such great wavelengths (the distance between two waves is about 12,000 miles), the oceans would have to be 14 miles deep to allow a free moving deep water wave to move at the earth's rotational speed at the equator (1050 miles per hour). The average depth of the oceans is only 2.3 miles. This results in slowing of the tidal wave to only 430 miles per hour.

This headlong rush of energy brings the flood — this enormous falling mass draws down the ebb. It is fortunate that lunar gravitation can only cause one vertical foot of displacement, or who knows what might happen when this wave hits the continental coastline.

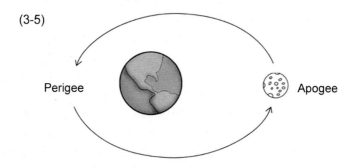

(3-5)

Lunar orbit is elliptical. Lunar gravitation is stronger at perigee and weaker at apogee.

In addition to the declination of the lunar orbit, there is another feature of the moon's orbit that influences the tides. The lunar orbit is not circular; it is elliptical, as in illustration 3-5. **The strength of the lunar gravitation is not always the same, because the moon is not always the same distance from the earth.** The point at which the moon is farthest from the earth is called the apogee. The nearest point is the perigee of the lunar orbit. The moon is about 31,000 miles closer to the earth at perigee. Like everything else in the business of celestial mechanics, the distance between the earth and moon is constantly changing in rhythmic cycles. Both gravitation and centrifugal force depend on this distance. Therefore, **the variation in lunar distance due to the moon's elliptical orbit is one of the factors contributing to the variation in the height of the tides throughout the weeks and months.**

The Lunar Orbit

The elliptical lunar orbit explains another rather subtle characteristic of the tide tables. The high tide mark changes almost every day. This change is rarely uniform or linear from day to day: it may increase two inches from Monday to Tuesday, increase three inches from Tuesday to Wednesday, and perhaps increase six inches from Saturday to Sunday, later that same week. We have already discussed how the moon's *distance* (and lunar gravitation at the earth's surface) changes due to its elliptical orbit. Now we learn that the moon's *speed* in orbit also changes. When the moon changes speed, *the rate of change* in the tides from day to day becomes nonlinear or non-uniform. Why does the moon change speed? Review illustration 3-5. As the moon moves from perigee toward apogee, it is traveling away from the earth, and slows down like a ball thrown upward. When the moon moves from apogee toward perigee, it is traveling toward the earth, and speeds up like a ball falling back down. **The rate of change in the earth's tides changes from day to day because the moon's orbital speed changes.**

The path of the lunar orbit also accounts for an important aspect of the tide tables familiar to all boaters: each day the high tide arrives about one hour later than the day before, and this is cumulative, so that if it is high tide at noon this Saturday, it will be low tide at noon next Saturday. Illustration 3-6 depicts the earth as seen from the north pole. **The essential reason for the daily change in the occurrence of the high tide (52 minutes later each 24 hours) is that the moon orbits the earth in the same direction as the west-to-east rotation of the earth on its polar axis.** Here is the explanation.

The definition of 24 hours is the time required for a point on the earth, such as S(1), to revolve 360 degrees around to the same alignment toward the sun, S(2). During this same 24 hours, another point on the earth, M(1), has also revolved 360 degrees. However, the moon is orbiting the earth every 27.5 days, and has moved from L(1) to a new location, L(2). M(1) must revolve greater than 360 degrees to align with this new lunar position, L(2). Therefore, the earth will have to revolve for an additional 52 minutes time before M(2) is aligned toward L(2), bringing the high tide 52 minutes later than the previous day.

(3-6)

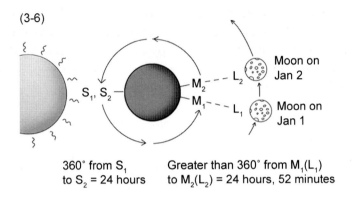

360° from S_1 to S_2 = 24 hours

Greater than 360° from $M_1(L_1)$ to $M_2(L_2)$ = 24 hours, 52 minutes

The shift from high tide at noon this Saturday to low tide at noon next Saturday is really just a mathematical coincidence. 52 minutes is about 5/6th of an hour. (5/6th hour shift per day) x (7 days) = 35/6 hour shift per week. 35/6 hours is only 10 minutes less that 36/6 hours, or 6 hours per week. 6 hours shifts high tide to low tide — a very handy coincidence. Of course, since our calendar is based on the phases of the moon, this is not entirely a coincidence.[c]

[c] Because our calendar has a common origin with the science of astronomy, there are many coincidences of tidal events with dates in history and religion. The famous "maelstrom" whirlpool, which instills terror in the mariners of the north Atlantic, and which inspired literature by Edgar Allen Poe, is located In the Saltfjord channel of northern Norway. Twice a day, an intense tidal current surges through a narrow rocky pass between this fjord and the ocean, causing a deep violent maelstrom, thirty feet wide.

It has always seemed significant to the local churchmen that the maelstrom is largest and deepest on Good Friday. They attributed this annual phenomenon with some mystical significance, as the maelstrom grows more angry at the date of the crucifixion. However, the event is less mystical when you are reminded that the Christian church established the date of Easter as the first Sunday after the full moon, following the vernal equinox. By definition, the tide generating forces of the sun and moon benefit from both the declination of the sun and a positive alignment of the moon and sun on this date.

As we expand the list of natural forces which influence the tides, we will be required to consider how these many forces interact with each other. For example, the declination of the lunar orbit causes the semidiurnal tides to be unequal height; and the elliptical orbit of the moon causes the lunar gravitation to be stronger at some times than at others. These two facts are totally independent. The angle of the moon's orbit to the equator has nothing to do with the elliptical shape of the orbit. Since they both influence the tide simultaneously, we must consider how these forces interact to determine the tide pattern.

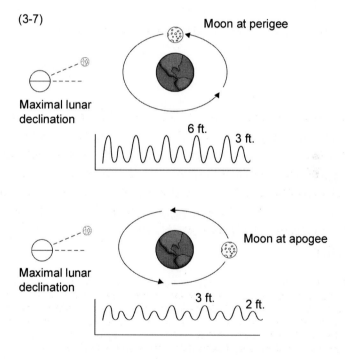

In illustration 3-7, there is an inequality of the semidiurnal high tides due to the declination of the lunar orbit. At the perigee of the orbit, lunar gravitation has a greater influence on the tides than it does at the apogee of the orbit. Therefore, at times of maximal declination, when the moon is also closest to the earth, there will be a larger difference in the semidiurnal tides. When the moon is farthest from the earth, there will be less difference in the tides due to the same amount of declination. This may seem to be a rather trivial point. It is included here as an example of how to organize your thoughts about the tide, as an exercise in how to analyze the tide tables.

The most fundamental concept in the entire process of calculating tide tables is that multiple independent forces influence the tide simultaneously; and both the height and the pattern of the tide depend on whether these forces reinforce each other, or interfere with each other.

We must dispense with one last feature of the lunar orbit and the phases of the moon, as we end this chapter. This may seem a little tedious, but it is necessary in order to understand the terminology used on the following pages.

A day is the time required for the earth to revolve 360 degrees on its polar axis. A year is the time required for the earth to orbit the sun. What is a month? That is not so simple. See illustration 3-8, below. Our calendar is based on 365 days per year. A *calendar month* is 1/12 of a year, or 30.25 days. This has no real relationship to the tide cycles.

A *sidereal month* (27.32 days) is the time required for the moon to orbit the earth 360 degrees, from perigee to perigee. Because the earth is revolving throughout the sidereal month, there is a difference in the time required for the moon to return to a position over the exact same longitude on earth. One lunar orbit plus this correction to reach the same longitude is called an *anomalistic month* (27.5 days). See illustration 3-8. This is directly linked to the daily tide cycle on earth, and is the period of a principle lunar tidal influence, which impresses a semidiurnal tidal wave on the oceans.

A *synodic month* is the time required between a new moon and the next new moon, or 29.5 days. This is defined by the way the moon appears to us (full disc, shadowed disc, crescent). Therefore this synodic month does not impress a daily tidal wave on the oceans. Instead, the synodic month (the period of 29.5 days between new moons) is linked to the fortnightly (biweekly) pattern of spring tides and neap tides that we will examine in the following chapter.

(3-8)

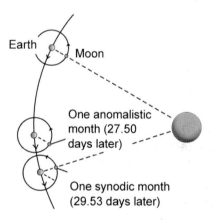

An anomalistic month is dependent on the orbit of the moon around the earth. This is directly linked to the daily tides.

A synodic month is defined by the appearance of the moon from the earth (the time between one full moon and the next full moon). This is not directly linked to the daily tides.

One anomalistic month (27.50 days later)

One synodic month (29.53 days later)

The earliest mariners noticed the very important relationship between the range of the tides and the moon's phases: full moon, quarter moon, new moon. Because the moon's phases and the spring tides and neap tides cannot be explained without including the tidal influence of the *sun*, we will elaborate the cause of spring tides and neap tides in the following chapter on solar influences.

Chapter Four

The Solar Influences and Solar–Lunar Interaction

"A scientist in his laboratory is not only a technician, he is also a child placed before natural phenomena, which impress him like a fairy tale."
Marie Curie

"Even if I could be Shakespeare, I would rather be Faraday."[a]
H. G. Wells

Charles Darwin once said that the discovery of a scientific fact was attended with prayer-like gratitude for being the first mortal permitted to view a portion of creation.[b] When Isaac Newton first formulated the Law of Gravitation, he immediately sought to verify his equation with scientific measurements of the lunar orbit around the earth. He based his calculations on the known size of the earth in the sixteenth century. When these calculations produced a result that conflicted with his theory, he did the honorable thing and reported that his theory had failed the test. Hugely disappointed, Newton put gravity aside and turned his attention to other matters for the next six years.

---

[a] Micheal Faraday was a brilliant English physicist. Many consider him the greatest experimentalist who ever lived. His experiments in electricity and magnetism laid the foundation for modern physics, and led to the invention of the electric generator and the electric motor.

[b] It may seem ironic that Charles Darwin, whose theory of evolution challenged biblical creation, would express his gratitude in this way. This reflects the modern intellectual error that science and religion are mutually exclusive. However, consider Bertrand Russell's opinion: "Religion tells us why creation was made, and science tells us how it was done," and Albert Einstein's opinion: "Science without religion is lame, and religion without science is blind."

Then, a respected contemporary scientist announced that the size of the earth had been revised (about sixteen per-cent larger). On learning this, Newton rushed back to his papers on gravitation, in order to repeat the calculations testing his theory. His biographers report that the great man was so excited that he was unable to do the math. He had to enlist an associate to calculate the lunar orbit, using the new data. This time the numbers perfectly matched his theory, which is now the Law of Gravitation.

For millennia, early civilizations all over the world understood that the phases of the moon were related to the height and range of the tide. In 1650, Bernhardus Varenius described this mystery: "...most philosophers, who have observed the harmony that these tides have with the moon, have given their opinion that they are entirely owing to the influence of that luminary. But, the question is, what is this influence? To which they only answer, that it is an occult quality, or sympathy, whereby the moon attracts all moist bodies. But these are only words, and they signify no more than that the moon does this by some means or other, but they do not know how: which is the thing we want." This is the context of ignorance and frustration that Newton exploded when his calculations proved right.

During full moons and new moons the tides are higher than during quarter moons. Anyone with an interest in this book is doubtless aware that there is a monthly cycle of higher "spring tides," during new moons and full moons, alternating with lower "neap tides," during quarter moons. The term spring tide has nothing to do with the spring season. Spring tide derives from the Saxon word "sprungen," meaning "to jump up." Neap tide derives from an old Scandinavian word meaning "hardly enough." **The lunar phases have nothing to do with the distance of the moon from the earth**. The appearance of the moon at different phases (full disc, shadowed disc, crescent) is just an indication of the alignment of the sun and moon to an observer on the earth. See illustration 4-1(a). When the sun and moon are aligned roughly in a straight line[c] which includes the earth, we will either see a full moon (the lighted portion) or a new moon (the shadowed portion). Whether we see light or shadow depends on whether the moon is located away from the sun, or toward the sun.

(4-1) (a) The blue bulges in the earth's oceans are caused by the lunar influences.

The yellow bulges are due to the solar influences on the tides. During full moons and new moons, the lunar and solar tides combine into the tidal wave.

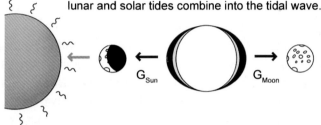

(4-1) (b) Just as in the previous illustration, the blue bulges are lunar, and the yellow bulges are solar in origin.

However, during quarter moons, these two tidal waves do not add together. Instead, the solar bulges are subtracted from the volume of the high tide aligned with the moon.

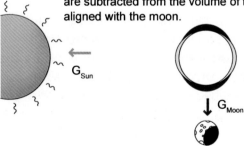

[c] When the sun, moon, and earth are perfectly aligned in a straight line, there is a solar or lunar eclipse. New moons occur when the moon is almost aligned directly toward the sun. Full moons occur when the moon is almost directly away from the sun, relative to earth.

When the sun and moon are aligned at 90 degrees or right angles to the earth, sunlight will cover half of the lunar surface, the other half will be in shadow; and we will see a quarter moon. See illustration 4-1(b). Again, none of this has anything to do with the distance of the moon from the earth.

During full moons and new moons, both lunar gravitation and solar gravitation are aligned to pull in the same direction, causing higher tides. During quarter moons, lunar gravitation and solar gravitation pull in different directions and the tide will have less range.

At the earth's surface, solar gravitation is about 50% as strong as lunar gravitation. Although the sun is 27 million times more massive than the moon, it is also 93 million miles away. Since the moon is only about one-quarter million miles from earth, lunar gravitation predominates, and the lunar tides are twice as great as solar tides.[d]

There are two bulges impressed on the earth's oceans by the solar tidal influence: a bulge due to solar gravitation on the side toward the sun, and a bulge due to the centrifugal force of the earth's orbit around the sun, on the opposite side. The solar and lunar bulges are created simultaneously, every minute of every day. Since there is only so much water to go around, so to speak, **the dominant lunar high tides may be enhanced by the solar bulges at spring tides, or diminished by the solar bulges at neap tides.** During spring tides, the high tide is the combined

---

[d] In a footnote to chapter two, we learned that (a) the *difference* in lunar gravitation on the opposite sides of the earth is the sole reason for the lunar tidal influence, and (b) this lunar net gravitational effect is inversely proportional to the cube of the lunar distance. The solar net gravitational effect is likewise due to the difference in solar gravitation on opposite sides of the earth. Therefore, the relative lunar/solar tidal influence is expressed mathematically as:

(2G × lunar mass × earth radius / lunar distance^3) / (2G × solar mass × earth radius / solar distance^3) = lunar mass × solar distance^3 / solar mass × lunar distance^3 = 2.3.

solar plus lunar bulge. During neap tides, there are separate solar bulges and lunar bulges, but since the lunar tide bulges are always greatest, we regard the lunar tides as the high tides for that twenty-four hours.

For those readers who like numbers, we can express this interaction as follows. At the earth's surface, solar gravitation is almost half as strong as lunar gravitation, very close to 5/11ths. Therefore, during all phases of the moon, the lunar contribution is 11 units of tide, and the solar contribution is 5 units. *The only question is whether they are making their contributions to the same oceans, or different oceans.* During full and new moons, they are aligned over the same oceans, and the height of the water at high tide is some multiple of 11 plus 5. During quarter moons the height of the water at high tide is some multiple of 11 minus 5. All other things being equal, the difference in the range of the tide between spring and neap tide will be: (11 + 5) / (11 − 5), or 16 / 6, or 8 / 3. Since it turns out that all other things are *not* equal, the difference between the tidal range during spring tide and neap tide is slightly less than 8/3, but this is pretty close.

(4-2) (a)

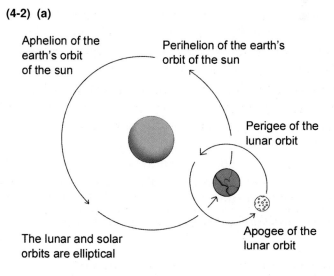

The influence of solar gravitation on the tide changes in two additional rhythmic cycles. See illustration 4-2 (a). First, the earth's orbit around the sun is elliptical, so that the distance to the sun varies. The distance to the sun is 3.4 million miles less at the perihelion (closest point) compared to the aphelion (farthest point). Obviously, **because of the earth's elliptical solar orbit, the strength of the solar gravitational field at the surface of the earth is constantly changing. This adds still another non-uniform constituent to the tides on earth and results in varying height of the solar tidal waves.**

The earth's elliptical orbit around the sun causes another influence on the tides. The *rate of change* of both solar and lunar influence on the tides is related to the speed of the earth and moon in their orbits. In chapter three we discussed how the elliptical lunar orbit causes the moon to change orbital speed. In this same way, as the earth travels from perihelion to aphelion, it is going away from the sun and slows down; and as the earth orbits back toward the perihelion, it is going toward the sun and speeds back up. Likewise, the speed of the moon in orbit is related to the distance to the sun because solar gravitation attracts the earth *and* the moon. Here we have an excellent example of how a change in one tide constituent (distance to the sun) can change other tide constituents (speed of the earth, *and* speed of the moon). This is another example of the non-uniform and nonlinear interaction of the tide constituents.

Secondly, the earth's polar axis is tilted 23.5 degrees from the plane of the solar orbit. See illustration 4-2(b). This has profound influence on the solar tidal wave and is essential to understanding tide patterns around the globe. It may be helpful to review the changing lunar declination shown in illustration 3-3, and compare it with the solar declination in illustration 4-2(b).

(4-2) (b)

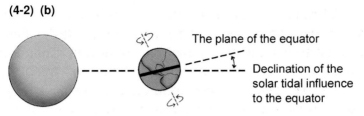

Tilt on polar axis causes seasonal variation in the tides

The solar declination differs from the lunar declination in that: (1) The two bulges of the solar tidal ellipse (solar tidal waves) are only about one-half the height of the lunar tidal waves. One of these small solar tidal waves is displaced 23.5 degrees beyond the equator due to the tilt of the earth's polar axis. The displacement of this smaller wave that far beyond the equator may *render it completely obscure in your hemisphere.* (2) Whereas the lunar declination crosses back and forth over the equator every two weeks, solar declination persists over the same hemisphere for six months, which allows one of the daily solar tidal waves to *establish a strong rhythmic influence within the oceans of one hemisphere.* Therefore, **the solar declination to the equator results in a *diurnal* solar tide pattern within each hemisphere.**

The lunar declination does not usually result in a diurnal pattern, because: (1) the lunar tides are over twice as high as the solar tides, and both lunar semidiurnal waves usually have significant influence over both hemispheres. Some of the complexities of this influence are discussed on page 67. (2) The lunar declination fluctuates north and south of the equator every two weeks, which does not allow a strong pattern of diurnal oscillation to develop in either hemisphere. Therefore, in spite of the lunar declination, **a semidiurnal pattern dominates the lunar tides over most of the globe**.

Just as summer, fall, winter, and spring are caused by the tilt of the polar axis, this results in a seasonal fluctuation in the solar tidal influence. **The solar declination has greater influence on the height of the water than the solar distance. At each point on the globe, the solar tidal influence is greatest when the sun is aligned over that hemisphere.**

In addition to the changing solar gravitation on the surface of the earth, we must also consider the changing solar thermal radiation. Unlike the moon, the sun radiates heat. The thermal influence on the tides depends on both: (1) the distance from the sun, and (2) the tilt of the earth's axis to the solar orbit, which brings summer or winter to each hemisphere. The oceans change average temperature seasonally, and this raises the sea level as the water heats up and expands. This **change in average sea level due to thermal expansion is a very significant influence on the height of the water. It raises the sea level by as much as 6 inches during the summer at the tropical latitudes, almost as much as the influence of solar gravitation**. This thermal effect is less important as you proceed toward the polar latitudes. This seasonal change in sea level due to the heating and cooling of the ocean is the only meteorological component included in the computer program for the tide tables at NOAA. Properly speaking, this is related to "climate," not weather. Thus, it is still accurate to say that the tide tables do not account for the weather.

These solar variables cause the phenomenon of extreme high tides that occur at many locations during certain months. Twice each year, there is a cluster of high tides that are almost twenty percent higher than the annual average high tide. The timing of this event depends on a coincidence of (a) the perigee of the lunar orbit, (b) spring tide solar-lunar alignment, and (c) favorable solar declination. It does not

The Solar Influences

correlate with the perihelion of the solar orbit, because solar declination has a greater influence than solar distance. Remember that favorable solar declination aligns the sun over our hemisphere, bringing summer and thermal expansion of the oceans.

The variation in solar tidal influence occurs on the scale of *months*, rather than days or weeks. At some point in your education on the tides, you will want look at a tide table and reckon for yourself which variations are due to solar influences, and which are due to lunar influences. You can do this already. Find a tide table that includes the entire year at a glance. This allows you to detect many undulating patterns of tidal fluctuation. You can now appreciate that when fluctuation occurs over a period of months, it is usually due to a solar influence. When fluctuation occurs over a period of days or weeks, it is usually due to a lunar influence. In other words, seasonal fluctuation is a sort of "solar signature." There are also "lunar signatures," such as tides that shift by 52 minutes each day. The tide table below is an example of the "fortnightly fluctuation" that we associate with full moons, new moons, and quarter moons, another lunar signature.

(4-3)

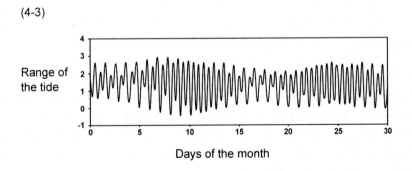

Before we end our discussion of the astronomical influences on the tides, we should take one last look at the effect of gravitation on the earth's oceans. We have referred to the enormous potential energy created by the vertical displacement of the oceans. At the surface of the ocean, the gravitation of the earth is 11,700,000 times greater than the net lunar gravitational effect. Therefore, the gravitational effect that causes the tides is much too weak to *lift* the oceans 12 inches vertically away from the earth. It is possible, however, to move the oceans horizontally within the earth's gravitational field. This gathers the oceans toward two points where the height of the water becomes elevated by the converging volume of water. Now, as we end our discussion of lunar and solar gravitation, we learn that **the major gravitational influence on the oceans is a *horizontal* vector of force, gathering the water around the surface of the globe toward the moon and sun, rather than a vertical lift of the water**.

(4-4) (a)

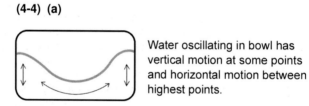

Water oscillating in bowl has vertical motion at some points and horizontal motion between highest points.

Illustration 4-4(a) depicts the oscillation of water in a bowl. If you started the water oscillating, and then added a single drop of blue ink at the *edge* of the bowl, you would see the motion was mostly vertical, bobbing up and down. If you added a drop of ink in the *center*, you would see the motion was mostly horizontal. The horizontal motion would be greatest between the two highest points on the surface.

Illustration 4-4(b) shows the earth with its lunar tidal bulges (labeled z). Just like the oscillating water in the bowl, the displacement of the water *is* vertical at the bulges, but **there is even greater horizontal motion of the water over a greater area of the earth's surface, between the bulges**, shown here as arrows. These horizontal vectors of force are called "tractal forces." We should now begin to have a concept of gravitation and centrifugal force gathering the oceans around the globe toward the maximal height of the tidal ellipse. This concept is necessary in order to explain the tides approaching the *western* coast of the continents. If your vision of the tides is limited to a vertical bulge in the oceans moving around the globe from east to west, it is difficult to explain the incoming tide on the coast of California, for example.

(4-4) (b)

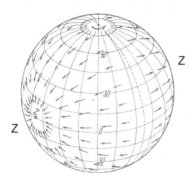

Tractal forces:

Although there are two points on earth with a purely vertical displacement of the oceans (point Z), most of the displacement is horizontal over the surface of the oceans (arrows). This horizontal vector is called a tractal force. This gathers the oceans toward the points of maximal vertical gravitational and centrifugal force.

The concept of a *horizontal* vector of force gathering the oceans toward the moon eliminates this confusion. For instance, while the moon is still over Kansas, the water in the Pacific Ocean is moving horizontally toward the moon, and up onto the beaches of California. There is minimal *vertical* lift of the water because the moon is over a continent (although the continent itself rises 4 to 8 inches higher above sea level). But, due to the tractal force, there is tidal movement toward the moon and toward both the east and west coasts. Thus, **the horizontal tractal forces gather the tides toward both the east and the west coasts of the continents.**

Yes, you *did* read in the previous paragraph that the continents are lifted up by lunar gravitation when the moon is over the land. The lunar and solar gravitation deform the shape of the earth, and raise the continents by as much as 8 inches, at the times corresponding to the highest annual high tides. In a similar manner, boats float higher in the water at high tide than they do at low tide, as the lunar gravitation lifts them up away from the earth. The sea level is slightly elevated along the coast of Chile, because the gravitational attraction of the Andes mountains pulls the ocean up and toward the coast. There is a "tide" in your body, which causes you to weigh less during high tides compared to low tides (about equal to the weight of several drops of sweat). Finally there are reliable reports of medical patients whose symptoms fluctuate with the tides. Before modern metal prostheses were available, one patient lost a large fragment of his skull, and lived out his life with his brain just below the scalp. At the most extreme high tides of each year, he suffered terrible headaches, which were relieved at low tide. This was confirmed by isolating him from any knowledge of the tides and matching his symptoms with the tide tables.

If *you* want a headache, consider this. The earth constantly changes shape as the moon orbits around it. Lunar gravitation deforms the surface of the earth, so that the earth is slightly thicker in the direction toward the moon. This occurs at the same point on the earth that is under the high tide influence of the moon, *lifting the ocean up*. However, the earth is now thicker at that same point, and this increases the earth's gravitation and *pulls the ocean down*, more than it would otherwise. Thus, the same lunar gravitation that raises the ocean changes the earth's gravitation so as to lower the ocean. We can understand everything we need to know about the tides without even knowing about this detail. It is included to give readers a sense of how thoroughly astronomy, geography, and oceanography are interrelated.

In the following chapters, we will turn our attention away from astronomy of the near solar system and the celestial influences on the tides. The remaining tidal influences are earthbound forces related to oceanography, hydraulics over the continental shelf, and geography. This is a good time to stop and note one of the central facts in our new understanding of the tides and tide tables: **Variations in the height and pattern of the tides that are seen on *different dates* at the *same coastal location* are due to astronomical influences: the motion and gravitation of the earth, moon and sun. Variations in the height and pattern of the tides on the *same date* at *different coastal locations* are due to earthbound influences: the dynamic oceans, the hydraulics over the continental shelf, and the geography of the coasts.** For example:

(1) Daytona Beach, Florida
High tide at 10 a.m., on January 15
High tide at 4 p.m., on January 22   Astronomical influence.

(2) Daytona Beach, Florida
Tidal range of 6 ft., on March 10
Tidal range of 4 ft., on June 20   Astronomical influence.

(3) Different shores in Florida
High tide at 6 a.m., at Miami on April 15
High tide at 9 a.m., at Mayport on April 15   Earthbound influence.

(4) Different shores in Florida
Tidal range of 3 ft., at Miami on July 4
Tidal range of 6 ft., at Mayport on July 4   Earthbound influence.

These earthbound influences (oceanographic, hydraulic, and geographic) are the explanation for the myriad patterns of tides that the British Admiralty found on the far-flung shores around the globe, the infinite variety of tides that could not be explained by the Equilibrium Theory of the Tides used by Isaac Newton.

Chapter Five

Celestial Harmonics

"There is geometry in the humming of the strings...
there is music in the spacing of the spheres."   Pythagoras

"The heavenly motions...are nothing but a song...
perceived not by the ear,
but by the intellect."   Johannes Kepler

In chapters one through four we have discussed the "celestial mechanics" that determine the changing influence of gravitation and centrifugal force on the earth's oceans. The motion in the near solar system is periodic, cyclical, and rhythmic. Its effect on the ocean is waveform. Long before this graceful rhythm was reduced to a computer program at NOAA, such natural wonders were once appreciated as "the music of the spheres."

Pythagoras, the father of geometry, was also fascinated with music and musical harmony. He discovered that the pitch of a musical note was the result of the length of the string that produced it. He then found that higher or lower octaves of the same note were produced by changing the length of the string by even multiples of the number two (half as long, twice as long). Next he explored the possibility that there might also be a numerical relationship between the "intervals" that divide the octaves into pleasing or "consonant" music. We commonly refer to these intervals as "do, re, mi, fa, so, la, ti, do." There are many other ways to divide the octaves into intervals, but this produces "dissonant" music that is not as pleasing to us. When Pythagoras discovered that there were exact mathematical ratios between the intervals that produce pleasing musical harmonies, it was revolutionary. It is not too much to say that he had invented western science. As biographer David Plant puts it, "By associating musical tones with measurements of length, he made the first known reduction of a quality (sound) into a quantity (length and ratio). The understanding of nature through mathematics remains a primary objective of science today."

This elegant and accurate description of nature inspired Copernicus to reject the accepted theory of the cosmos in the 16th century. His measurements of the planetary orbits did not correlate with the ancient Ptolemaic system, and he sought numerical relationships between the distances of the planets orbiting the sun. Unfortunately, when he rearranged the furniture in the solar system, the pope tripped over this new arrangement — in the dark, you might say.

A hundred years later, Johannes Kepler vindicated poor Copernicus. By careful scientific observation and creative intellect, he described the workings of the solar system, overcame the resistance of the church, and altered western civilization. In order to accomplish this, he employed the same scientific method that Pythagoras first used to investigate musical harmony.

To anyone who thinks it is hyperbole to say that these men, "altered western civilization," I offer the following example. In the middle ages, many people believed that beyond the clouds and the moon there was a solid black roof over the sky. This was the outer limit of the natural world where they lived on "the firmament." Outside the roof of the firmament was everywhere a brilliant white light. The stars where actually tiny holes in the roof of the firmament. If this was so, the starlight allowed them to peek outside the natural world, and get a glimpse of heaven itself. This view was completely wrong, but wonderfully inspirational. This may sound foolish to you. If so, I highly recommend that you go outside on the next clear night, and look at the night sky. Visualize the stars as holes in the roof of the firmament, and imagine you are seeing light from outside the natural world. Try this, and *then* decide what you think of this notion.

Now, we live in the post-lunar-landing era. We know that the stars are actually masses of hydrogen collapsing under their own gravitation, and the starlight comes from thermonuclear fusion, the stuff of hydrogen bombs. Quite a different viewpoint. Science has replaced mystery with certainty. There is more knowledge in the world and less inspiration.

We have now measured the orbits of the solar system so precisely we can list the period (time elapsed between cycles) of every celestial cycle that influences the tides. The following table is only a partial list of these tide-generating forces.

Partial List of the Astronomic Cycles Related to the Tides

Phenomenon	Period	Cause
Semidiurnal tide	12.43 hours	Rotation of the earth relative to the moon
Diurnal tide	23.93 hours	Influence of lunar and solar declination
Lunar fortnightly effect	13.66 days	Varying lunar declination
Spring tide interval	14.75 days	Phase relation between the sun and moon
Anomalistic month effect	27.55 days	Moon's elliptical orbit of the earth
Solar semiannual effect	182.6 days	Varying solar declination
Anomalistic year effect	365.26 days	Earth's elliptical orbit
Regression of the moon's nodes	18.6 years	Period of the maximal amount of lunar declination

Each orbit in the near solar system, each cycle with its own period of minutes, hours, days, and years, shapes the pattern of the tides on earth. Of course they all influence the tides simultaneously. That is, every orbit and cycle is making its contribution to the tidal wave[a] at the same time, all of the time. In order to make sense of such a complex system, we need to review the basic physics of wave interaction: the concepts of *wave reinforcement* and *wave interference*. See illustration 5-1, below.

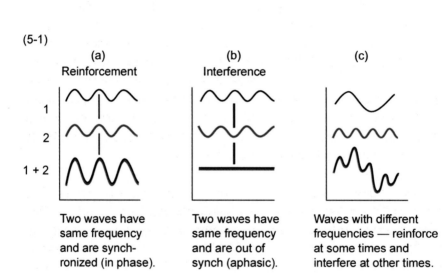

(5-1)

(a) Reinforcement — Two waves have same frequency and are synchronized (in phase).

(b) Interference — Two waves have same frequency and are out of synch (aphasic).

(c) Waves with different frequencies — reinforce at some times and interfere at other times.

---

[a] The term "tidal wave" is commonly misused in reference to those rare natural disasters that inspire Hollywood movies, the unpredictable and catastrophic wall of water that assaults the coast, causing death and destruction. In fact, these are properly called "tsunami waves." Tsunami waves are the result of volcanic eruptions, earthquakes, or landslides on the sea floor. Tidal waves are the normal bulges in the ocean that circle the globe every day that we are discussing on these pages.

When the rhythmic celestial forces are depicted on a graph, they have an undulating waveform (~~~~~~~~~). When two waveforms are superimposed, they may: (a) reinforce each other if they are synchronized with each other (in phase), or they may (b) interfere with each other if they are "out of synch" (aphasic). In illustration 5-1(a), the waves are the same frequency and in phase. Note how they reinforce each other, and create a waveform of greater amplitude. In illustration 5-1(b), the waves are the same frequency and perfectly aphasic. Now they completely cancel each other out. If the waves have different frequencies, as in 5-1(c), they will reinforce each other at some times, and interfere with each other at other times. **The concept of reinforcement and interference of multiple, simultaneous, and independent forces of nature is one of the most fundamental principles in this book. The current scientific theory of the tides and the computer model used to construct the tide tables are based on this principle.**

Look again at our partial list of the astronomical cycles that influence the tides. Remember that **each orbit and cycle in the near solar system impresses a wave on the ocean**. There are only three celestial bodies involved: the sun, moon, and earth. However, the complex interaction of their orbital speed, distance, and direction, and the changing declination of the sun and moon to the equator, **all of this translates into multiple waves of tidal energy circling the globe**. There is a tidal wave that repeats every 12 hours, 26 minutes, due to the rotation of the earth relative to the moon. Another wave of tidal energy repeats every 24 hours (actually, 23.93 hours), due to the changing declination of the lunar and solar orbits to the equator (review page 51). There are many other tidal waves circling westward through the oceans.

In the terminology of illustration 5-1, **these tidal waves are sometimes in phase, and sometimes aphasic, and they all have different frequencies. Therefore, they will reinforce each other at some times, and interfere with each other at other times.** Once you realize that all of these waves are constantly making a contribution to the height of the water, constantly reinforcing each other or interfering with each other, you begin to get an idea of why the tides have different patterns on different days at the same point on the coast.

A friend of mine returned from a fishing trip to the west coast of Florida with a question about the curious tide pattern he had seen there. On Monday there were two high tides and two low tides in 24 hours, but on Saturday of the same week there was only one high tide and one low tide. He had asked the local fishing guides for an explanation. They made their living on these waters, but they couldn't explain this mixed tide pattern.

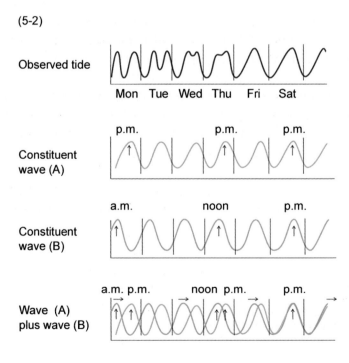

(5-2)

Nature is awesome in complexity and at the same time beautiful in simplicity. The explanation of the mixed tide pattern in southwest Florida is an excellent example of this paradoxical quality of nature. In order to understand how such unlikely tide patterns occur, we can use a simplified model with only two hypothetical tide constituents. See illustration 5-2.

Notice that the high tides on our hypothetical coastline are actually the crests of two different tidal waves, labeled (a) and (b). On Monday, these two crests occur at different times of the day, and there are two high tides. Notice that wave (a) occurs every afternoon of the week. Wave (b) has a longer frequency than wave (a). Therefore, every twenty-four hours, wave (b) crests later in the day. Wave (b) occurs in the morning on Monday, at noon on Thursday, and in the afternoon on Saturday. Each successive day, the crest of wave (b) moves closer and closer to the afternoon crest of wave (a). By Saturday, the two tidal waves crest at the same time; they are superimposed, and create one high tide. If you project this process into the following week, you will find that wave (b) crests *later* than wave (a). Once they become separated, there will be two high tides again.[b]

The seemingly chaotic and complex patterns of the tides are actually very logical and simple, once you analyze the interaction of their constituents. Here we have a great lesson about the tides and nature in general.

We have used a hypothetical example, with two tidal wave frequencies, to demonstrate how reinforcement and interference

---

[b]There is another explanation for the changing tide pattern shown in illustration 5-2. It could be due to the changing declination of the lunar orbit to the equator. Under the right circumstances (the right latitude, the right time of month, etc.), both of the semidiurnal tidal waves might be manifest on the same side of the equator on Monday; and then on Saturday, one of these semidiurnal waves might be declined farther below the equator, away from the latitude of our coast, leaving only one recognizable high tide.

of waveforms can explain unusual tide patterns. If you have a skeptical nature (which is the nature of every good scientist), you may want the specific details of the strange southwest Florida tide pattern. By the end of chapter nine, you will see that there are many simultaneous influences on the tides in this part of the world, and it would surprise us if these tides were *not* strangely unique and uniquely strange.

(5-3)

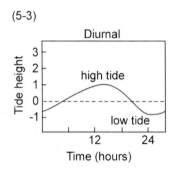

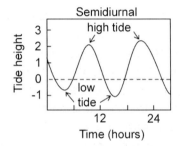

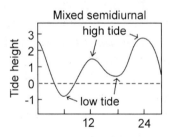

"Mixed diurnal" tides usually have only one recognizable high tide each 24 hours. They are different from diurnal tides in that the hidden semidiurnal component occasionally becomes evident. Both "mixed, diurnal" and "mixed semidiurnal" tide patterns are extremely erratic and variable on different dates, at the same port.

All of the various tide patterns on earth are classified into three types. (1) *Semidiurnal tides* consist of two high tides and two low tides each day, which are about equal height. Each tide cycle consists of about 6 hours of flood tide and 6 hours of ebb tide. This is the most common pattern of tides on earth. (2) *Diurnal tides* consist of one high tide and one low tide each day, flooding for about 12 hours and ebbing for 12 hours. This pattern is quite rare, but does occur in a few gulfs and seas around the globe. (3) *Mixed tides* may consist of two daily high tides with very unequal highs and lows, or only one high tide daily. Mixed tides usually have a widely changing pattern throughout the year. Mixed tides are divided into mixed-semidiurnal or mixed-diurnal, based on the predominant pattern. These three types of tide patterns are shown graphically in illustration 5-3.[c]

Although it is valid to state that mixed tide patterns (in fact, all tide patterns) are the result of the reinforcement and interference of multiple tidal waves, there is another model that is helpful in understanding these three different basic tide patterns. On pages 33 and 51, we discussed the effect of the declination of both lunar and solar orbits to the equator. One of these effects is to displace one of the semidiurnal tidal waves beyond the equator, so that it may be unaccounted for in the oceans of one hemisphere. Therefore, the degree of declination of these orbits to the equator may make a major contribution to the tide pattern. A small declination is associated with semidiurnal tides, since the crests of both semidiurnal waves will be at the same latitudes, and arrive at the same coast twice each day. Maximal declination may lead to diurnal tides, as one of the semidiurnal waves is declined beyond the equator. This is what we commonly see at the polar latitudes, for instance. Moderate declination often yields mixed tides, as certain tidal constituents may influence the tide pattern at some times, but not at others — their influence will change with successive tide cycles, as the declination is changing every day.

---

[c]For the technically oriented reader, tide patterns can be classified by a ratio of two *diurnal* tide constituents called K1 and O1, and two *semidiurnal* tide constituents called M2 and S2. The formula **is (K1 +O1) / (M2 + S2)**. If this ratio is less than 0.25, the tide is called semidiurnal; a ratio between 0.25 and 1.50 is called mixed-semidiurnal; a ratio between 1.50 and 3.00 is called mixed-diurnal; and, a ratio greater than 3.00 is called a diurnal tide.

In chapter seven, we will learn why different oceans, gulfs, and seas support semidiurnal tides better than diurnal tides, or vice versa; and why mixed tides occur in some oceans, but not others. For now, we understand that **all tides on earth include the contribution of every single celestial cycle, every astronomical influence. The constant reinforcement and interference of these multiple waves of energy impressed upon the oceans is a major reason for the variety of tide patterns seen along the coasts.**

This variety of tide patterns is most extreme in the oceans, seas, gulfs, and bays that have mixed tides. In illustration 5-4, the red line represents the actual measured height of the water during a 24-hour cycle, on three different days of the year, on the coast of California. These variations can be explained as the sum of the two most dominant celestial cycles: the lunar (black) semidiurnal tidal wave, and the solar (yellow) diurnal tidal wave.

(5-4)

A

B

C

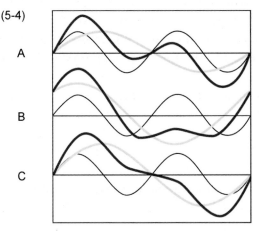

Red line = actual tide patterns on three different days, during three different months, on the coast of California.

Black line = semidiurnal tidal wave.

Yellow line = diurnal tidal wave.

Reinforcement and interference of black line plus yellow line = red line, the summation tidal wave, the measured height of the water during a tide cycle.

The black line (semidiurnal) is essentially the same on all three days. The only difference between day (A) and day (B) is that on day (B) the yellow diurnal wave has a slightly higher amplitude and is shifted to the left (earlier in the day). The only thing required to create the different pattern on day (C) is to shift this same diurnal wave to the right (later in the day) compared to day (B).

If you compare day (A) to day (C), the semidiurnal waves are the same, and the diurnal waves occur at the same time of day — the only difference is that the diurnal wave has slightly higher amplitude on day (C). Notice how a slight change in one constituent of the tidal wave has a dramatic effect on the tide table, once it interacts with other simultaneous constituents. In particular, notice that there are two high tides and two low tides on day (A), but only one high and low tide on the other days; and this is accomplished by very slight variations in the timing and the strength of one of the tidal constituents.

Furthermore, the movement of the tide is almost flat during the middle of day (C). It is easy to see that this period of about six hours could be made completely flat (a condition known as a "stand" in the tide) by the right combination of these constituents.

Considering all of the possible combinations of all of the many celestial cycles, it is obvious that **this finite list of orderly rhythmic tidal constituents can combine to produce the seemingly chaotic, infinite variety of tides on earth.**

In the following chapters on the oceanographic, hydraulic, and geographic influences on the tides, readers are reminded that we are further expanding the list of simultaneous, independent forces which are constantly interacting to generate the ultimate measured tide experienced by mariners out on the water. **It is the summation of dozens of influences (astronomical, oceanographic, hydraulic, and geographic), which may reinforce** each other or interfere with each other, that ultimately determines the height of the water during any hour of the tide cycle at each location on the coast. This principal is the key to understanding much of what follows in this book.

Chapter Six

The Coriolis Force and Oceanic Amphidromes. Coastal Kelvin Waves. Tidal Currents. Sea Level

"No matter what you look at, if you look closely enough, you are involved in the entire universe."　　　　Richard Feynman

We are schooled in the bias that our ancestors in early Europe built the foundations of modern science, that Euclid devised trigonometry, and Columbus discovered the world was round. The truth is that in Persia, around 830 A. D., an astronomer of the school of Al-ma'mun, who had no contact with Europe, not only knew the earth was a sphere, he also calculated the circumference of the globe within a 5% margin of error. How did he manage this feat without even leaving the desert?

Near the Euphrates river there was a deep well, a narrow pit that penetrated far into the earth. The sunlight reached the bottom and reflected off of the water for only a few minutes on one day of each year. About fifty miles away, in the neighboring village of Palmya, there was a ceremonial pole of great height. He reasoned that both the well and the pole pointed directly at the center of the earth. He had a young assistant measure the exact distance between them, and measure the height of the pole. Then, at the very moment that the sunlight reached the bottom of the well, he measured the length of the shadow from the pole. Using simple trigonometry, it was possible to calculate the radius of the sphere, and then the circumference of the earth.[a]

---

[a] Eratosthenes of Cyrene is also given credit for measuring the earth's circumference, around 200 B. C. His procedure was based on the position of the sun being directly over the town of Syene, at the same time that it was 7 degrees from zenith in Alexandria.

One of the major reasons that the tides are so complex is because they occur on a rotating sphere. To an observer looking down on the earth from the moon, the tide doesn't really revolve around the earth each day. It is the globe that revolves, while the tidal ellipse remains in fixed alignment with the moon. Because of the earth's rotation, additional forces influence the oceans and the coastal waters. These are known as "geostrophic forces."

Astronomical gravitation and momentum are the engines that set the tides in motion, but the influences that shape the tides into different patterns at the coast and determine the height of the water in your local harbor are earthbound: the rotation of the earth, the dynamics of the oceans, the shallow water hydraulics over the continental shelf, and the geography of the continents.

(6-1)

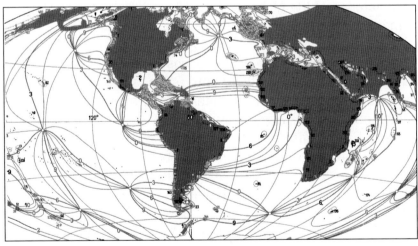

Beginning in the late 1800's, oceanographers began systematic measurement of the path of the tidal wave across the earth's oceans. They found that the tidal wave does not move directly from east to west around the globe. First they charted the arrival of the high tide at each point on the ocean's surface. Then they connected these points with lines called "co-tidal lines." Each co-tidal line connected the points where the high tide arrived simultaneously. They found that these lines were not straight across the oceans from north to south, with successive high tides moving directly from east to west. Instead, they were arranged in a circular pattern over the oceans. See illustration 6-1.

As you can see, these patterns are not perfectly circular, due to interruption by the continents. They named these circular systems of tidal wave movement "amphidromes," from a Greek word for racetrack. We will look more closely at these amphidromes after we discuss the cause of this unlikely finding.

(6-2)

(a) Pencil draws a straight line across a stationary sheet of graph paper.

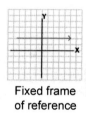

Fixed frame of reference

(b) Pencil moves in a straight line across a piece of graph paper that is turning 360 degrees counterclockwise.

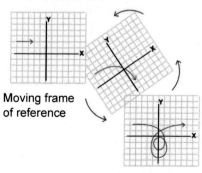

Moving frame of reference

First we must consider the important concept of a "frame of reference." The tidal wave does not move across a stationary ocean with a flat surface. It moves across the surface of a rotating sphere. See illustration 6-2.

If you take a pencil and move it directly from left to right across a graph, you will draw a straight line, as in 6-2(a). However if you make the same linear motion of the pencil across a graph that is turning counterclockwise while you draw, the pencil will trace out a clockwise circular pattern, as in 6-2(b). This does not occur because the path of the pencil was circular — it occurs because you were drawing on a moving "frame of reference." If an observer watched from across the room (from a stationary frame of reference), he would report that the pencil traveled in a straight line, while the graph paper moved. If an observer was standing on the graph paper (on a moving frame of reference), and if the paper stretched out to the horizon, he would report that the pencil moved in a circular path over the paper. No matter which observer says what, the fact is that there is a circular pencil line on the paper — we are not talking about an illusion here. The point is that this circular path is not because the pencil is moving in a circle. It is due to the motion of the paper, the moving frame of reference.

As the earth revolves on its polar axis, the tidal ellipse aligned with lunar and solar gravitation is drawn directly around the globe from east to west. And yet, measurements of the ocean reveal that the motion of the tidal wave describes a circular path. This is because the tidal wave must move over the surface of a rotating sphere (what physicist call a non-inertial frame of reference).

Gaspard Gustave de Coriolis was the first to understand that the paths of objects, air, and water moving over the surface of the earth were diverted into circular patterns by an effect now known as the Coriolis force. I suppose we should all be forever grateful that we do not have to call it the Gaspard Gustave de Coriolis force. This causes moving objects to be diverted to the right (clockwise) in the northern hemisphere, and to the left (counterclockwise) in the southern hemisphere.

(6-3)

(a)

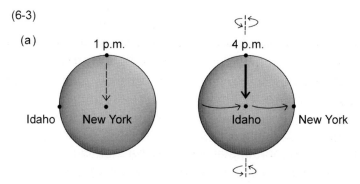

Seen by observer in outer space: airplane travels in straight line while earth rotates Idaho around to previous location of New York.

(b)

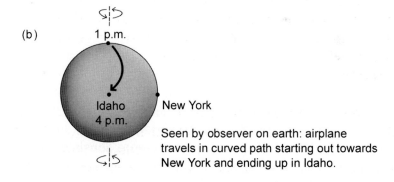

Seen by observer on earth: airplane travels in curved path starting out towards New York and ending up in Idaho.

Let's look at a simple example of how this force affects the path of an object moving over the earth's surface. Illustration 6-3 depicts the flight of an airplane leaving the north pole on autopilot course for New York. This flight departs at 1:00 p.m. and arrives at 4:00 p.m. During these three hours, the earth rotates the longitude of New York about two thousand miles to the east. Likewise, the state of Idaho is rotated to the east, and, in a sense, arrives at the previous location of New York. If the pilot maintained an absolutely straight course, the airplane would start out headed toward New York, and would end up in Idaho.

As shown in illustration 6-3(a), an observer in outer space would see the airplane travel in a straight line for three hours, while the earth rotated Idaho and New York toward the east. He could explain all of this by simply using Newton's Laws of Motion. However, an observer on the earth, not accounting for the earth's rotation, would see the airplane travel in a curved path turning clockwise over the surface of the earth, as in illustration 6-3(b). This observer on earth, standing on the non-inertial frame of reference, *could not explain what he saw, using Newton's Laws.* He would need to invoke some additional force acting on the airplane, causing it to curve to the right, in order to explain his observation — this could be perfectly satisfied by our Coriolis force. In fact, airline pilots must correct for this phenomenon on every flight that travels a significant distance, in order to navigate properly around the surface of the earth.

Artillery gunners are provided with sighting mechanisms that correct for the deflection of projectiles, which would otherwise land a significant distance from their intended targets due to the Coriolis force. During the naval engagement in the Falkland Islands during World War I, the gunners were consistently missing the German ships by 100 yards to the left, until someone realized that the Coriolis correction in the gun sights was calculated for warfare *north* of the equator. Apparently, they were under the impression that all naval battles would take place in the northern hemisphere.

In order to understand the Coriolis force, and why objects are diverted from a straight path into rotary motion as they move over the surface of the earth, we would have to depart far from oceanography, and travel pretty deep into the world of physics. Because this will doubtless appeal to readers with a scientific and mathematical background, there is a brief discussion of the conservation of angular momentum at the end of this chapter,[b] and there are ample resources listed in the bibliography.

This same force is the cause of the circular movement of atmospheric winds, which rotate around the centers of high and low barometric pressure. In fact, the major reason that these "highs" and "lows" develop is because of the Coriolis force. When the atmospheric pressure becomes greater at one location, the air begins to move toward nearby lower pressure. If these winds could continue in a straight line, the pressure would soon equalize, and the highs and lows would never develop. However, as soon as a mass of air begins moving, it is diverted into a rotary pattern by the Coriolis force. This prevents these winds from reaching their low pressure destination, and creates circular patterns of wind around transient centers of high and low pressure.

The Coriolis force diverts the path of moving objects (and masses of air or water) toward the right (clockwise), in the northern hemisphere, and toward the left (counterclockwise), in the southern hemisphere. However, the movement of the wind around "lows" is counterclockwise in the northern hemisphere. Illustration 6-4 provides the explanation for this apparent contradiction. Note that air masses converge toward the low pressure center from all directions. These winds are then diverted toward the right by the Coriolis force. However, they do not continue to circle back away from the "low" because of the pressure gradient that attracts them — they find a transient equilibrium, by moving tangential to the center of low pressure,

---

[b] See supplement to chapter six, at the end of this chapter.

and they join the other converging winds, now moving counterclockwise as a weather system. Hurricanes in the northern hemisphere exhibit this counterclockwise pattern. Tornadoes are governed by stronger local weather conditions, and are not related to the rotation of the earth, just as whirlpools are the result of intersecting local currents.

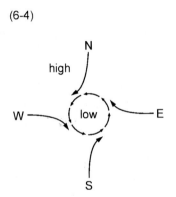

(6-4)

Note that each air mass (N, E, S, W) is diverted toward the right (clockwise) as it begins to move toward the low pressure center.

However, as they turn tangential to the "low," they converge into a counterclockwise pattern.

There is a transient balance of forces — they are still diverted to the right by the Coriolis force, but they are attracted toward low pressure.

This creates a counterclockwise weather system around the "low."

The great ocean currents tend to be diverted from a straight direction of travel, and assume roughly circular patterns, turning to the right in the northern hemisphere, and have the opposite rotation south of the equator. The path of the gulfstream through the Atlantic ocean and the Humbolt current in the Pacific are influenced by the earth's rotation.

I suspect that you are wondering why I have not used the example of water being flushed down the drain of a basin in your house. This appears to be the most common and recognizable manifestation of the Coriolis force. The everyday experience of seeing water empty down a drain in a counterclockwise whirlpool, in the northern hemisphere, seems a useful analogy to the Coriolis effect in other spheres. The counterclockwise rotation might seem analogous to the weather system in illustration 6-4, as the Coriolis force might divert the water to the right, while gravitation attracted the water toward the center of the drain. However, the effect of the Coriolis force depends on the speed of the motion of an object (or water) and the distance between the latitudes involved. Theoretically, the water moves too slowly and the distance across a basin in your home is too small to generate a Coriolis induced whirlpool within such a basin. Hydraulic engineers invoke other forces to explain this phenomenon, such as irregularities in the shape of such basins, convection currents in the water due to different temperatures across the basin, and residual water motion from the filling of the basin (imperceptible motion that persists long after filling the basin).

Under very carefully controlled laboratory conditions, with perfectly regular basins, zero initial motion, and uniform water temperature, the pattern expected from the Coriolis force *can* be demonstrated in such basins of water emptying down a drain. In the final analysis however, it is somewhat puzzling to scientists that the northern hemisphere-counterclockwise and southern hemisphere-clockwise pattern is seen in these small basins in which it is virtually impossible for the Coriolis force to have this effect. Perhaps there is a conspiracy among toilet bowl manufacturers in both hemispheres that causes this phenomenon.

By now, you must have guessed where this was leading. As the tidal wave moves over the revolving surface of the earth, it too is diverted into a pattern of circular motion around the oceanic amphidromes. Like the winds around atmospheric low pressure systems, the tidal wave moves counterclockwise around the amphidromes in the northern hemisphere, and in the opposite direction south of the equator. Although this too is linked to the earth's rotation, it is not so simple as our example of an airplane or the wind curving off course due to the rotation of the globe.

The complex mechanism that links the earth's rotation and the Coriolis force to the amphidromes involves the solution of advanced hydraulic dynamics, of which the Coriolis force is only one of many constituents. If you should decide to research the literature on this subject, you will find that there is considerable controversy about the mechanism of the oceanic amphidromes. Compared to astronomy and physics, oceanography is a relatively new branch of science. One oceanographer told me that it is impossible to convene ten oceanographers together in a room without a debate about, "whether the ocean currents cause the sea level to be higher at some locations, or whether the differences in sea level cause the ocean currents," or some similar circular argument (no pun intended). The technology necessary to begin the exploration of the ocean depths has only been available for a few decades.

The complexity of global ocean dynamics and the mechanism of the oceanic amphidromes includes the many gradients (slopes) in the ocean surface: gradients due the water pushed ahead of the great ocean currents, unequal sea level caused by the trade winds, and gradients due to the tidal ellipse itself — when it is high tide on one side of an ocean and low tide on the other side, still another slope is created. Water is constantly flowing down all of these gradients. Then, because some regions of the oceans may differ in temperature or salinity, additional horizontal gradients add further to the dynamics of the oceans. The oceans also have distinct vertical layers of differing temperature and salinity, and these layers often behave independently from the surface. This adds to the ocean dynamics by vertical sinking and upwelling on a global scale. Then, there is the infinite irregularity of the seafloor, including mountain ranges and immense canyons, and the extreme complexity of the shape of the continents and oceans. All of this is compounded by our Coriolis force.

For all of these reasons, the mechanism by which the oceanic amphidromes develop requires further investigation by oceanographers. In the meantime, we must be satisfied with the concept that **the tidal waves are moving over the surface of a rotating sphere, and this causes a circular pattern of oscillation on the surface of our oceans**. In connection with such unlikely findings, a wise man once said, "The fact that something *does* happen, proves that it *can* happen." Although it is somewhat difficult to explain how they come about, there is no doubt that the amphidromes are an important element of the tides on earth.

(6-5)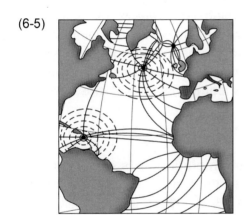

Co-tidal lines (solid) indicate the time of arrival of the high tide, moving around the amphidrome.

Co-range lines (broken) indicate the progressive increase in the range of the tide toward the outer edge of the amphidrome.

Now, let's consider how this affects the range and the pattern of the tides. The amphidromes can be considered as enormous oceanic whirlpools, thousands of miles across. See illustration 6-5. The water is lowest in the center of a whirlpool, and highest at the outer circumference of a whirlpool. In addition to the "co-tidal lines," which radiate like spokes of a wheel around the amphidromes, there are "co-range lines," which are concentric circles around the center. The range of the tidal wave is slightly higher as you proceed farther toward the outer edge of each amphidrome.

Along the coasts of the continents, the influence of the oceanic amphidromes is relatively minor compared to the other constituents of the tide pattern. At most coastal ports, there is usually some predominant influence that obscures the influence of the amphidromes. However there is one result of the amphidromes that is significant along many shores. It is puzzling to many experienced mariners that the time of arrival of high tide is different at the different ports along a shoreline. For example, there may be a pattern of later and later arrival of high tide as you proceed along a coast from north to south, or vice versa. In some places, this is because the high tide sweeps up or down the coastline, as part of the circular pattern of the tidal wave. **The location of the nearest amphidrome may influence the time of the high tide at each point on the coast. One reason that the tides sweep up or down a coastline is that the tidal wave motion is roughly circular.** Other causes of delayed tides are discussed in chapter eight.

Proximity to the nearest amphidrome is one more factor influencing the range of the tide at different ports around the globe. Those islands and those ports along the coast that are farthest from the center of the nearest amphidrome may have slightly higher tides due to this influence.   Unlike the continental coasts, mid-oceanic islands do not experience large tidal ranges. Hawaii, the Caribbean Islands, and Tahiti have tides of about one foot. These islands are near to the center of an amphidrome — the tidal elevation rotates around them, but at the center of this motion the water level is relatively unaffected by the tide. Another reason that most mid-oceanic islands have a relatively narrow tidal range is that they only experience the small gravitational tidal bulge of the oceans. Higher tides are only found on the coasts of the continents, as we shall see in chapter eight.

The Coriolis Force and Oceanic Amphidromes

The illustration below depicts the Hawaiian Islands at the bottom right, and the nearest Pacific amphidrome at the bottom left. The shades of grey indicate the increasing range of the tide as you progress farther from the center. The co-tidal lines show that high tide revolves around the amphidrome counterclockwise. This circular path of the tidal wave is the only reason for the fact that high tide arrives first at Hawaii, then at Maui, then at Oahu, and last at Kauai.

Just as in the Hawaiian Islands (shown below), the time of arrival of the tide on the coasts of the continents is related to this motion. It is just not as simple to discern this pattern on the continental coasts, because of other dominant influences on the tides.

(6-6)

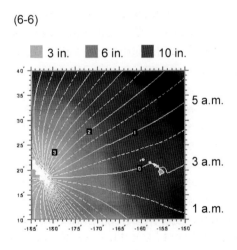

The amphidrome nearest to the Hawaiin islands is on the left. The co-tidal lines progress counterclockwise.

The island chain is on the right. In order from the bottom to the top are: Hawaii, Maui, Oahu, and Kauai.

The high tide arrives first at Hawaii, then Maui, then Oahu and last at Kauai.

The sole reason for the difference in the hour of the high tide at each island is the tidal motion around the amphidrome.

It is hard to imagine how this massive deformity of the ocean surface could be caused by something as transient as a single tidal wave. The major cycles of the tide pass across the ocean in 24 hours or less. However, the amphidromes represent an equilibrium established by the relentless forcing of energy through the oceans by many successive tides. Only after many repetitive tide cycles could this astronomical displacement of the vast oceans occur and create such an unlikely thing as the amphidromes.

The rotation of the globe causes another important influence on the height of the tides. This is a phenomenon known as a "coastal Kelvin wave" that occurs in water moving through a channel confined by a landmass. If you hold a teacup that is halfway full and swirl the fluid around with your hand, you can look down into the cup and see that the fluid is higher on one side and lower on the opposite side. On a global scale, the rotation of the earth impresses a force on water moving through a channel, and the water becomes inclined higher on one side than the other. Because of the earth's rotation, amphidromes develop in *open* basins such as the oceans, and coastal Kelvin waves redistribute the water in *confined* basins such as the English Channel.

If you could somehow view a cross section of a whirlpool or an amphidrome, you would see that it was lowest in the center and highest at the outer edges. If you viewed a cross section of a confined linear channel of water moving north, (in the northern hemisphere), you would see that it was higher on the east side of the channel, because the inertial force of water moving south-to-north presses against the right (east) bank in the northern hemisphere. This is a rather complicated hydraulic phenomenon, but an oversimplified explanation is that: (1) the amount of the west-to-east velocity due to the earth's rotation depends on the distance to the earth's rotational axis. (2) The latitudes closer to the equator are further from the earth's rotational axis. Therefore, a stream moving away from the equator has an excess of west-to-east momentum imported from the more equatorial latitudes. This excess momentum manifests itself as higher water on the east shore of such channels.

Within the English Channel, the tidal ranges are higher on the French side because of a coastal Kelvin wave that tilts the water level higher toward the east. This is not a subtle detail: it is a powerful force. Some eastern coasts of the English Channel have some of the most extreme tides on earth, being over 30 feet. One of the few tide-powered electric generating stations on earth is on the Rance estuary of the French coast. At another section of this coast, the incoming tide is so abrupt that surfers are able to ride the tidal wave up the Severn River. This amplification of the tide toward the eastern side of the English Channel is an excellent example of how **the earth's rotation generates coastal Kelvin waves, which amplify the tides at one side of channels which are partially confined by a landmass**.

Still another consequence of the earth's rotation is the pattern of tidal currents that we experience in coastal bays and harbors. The tides circle the globe as an "oscillatory wave," in which energy moves through standing water; the water itself does not have significant net horizontal motion as the tidal wave passes through. On the other hand, the tidal currents are moving water. The direction of the tidal currents is easily noticeable within large coastal basins. If there are many sailboats moored out in a harbor on windless days, the alignment of the boats will only reflect the direction of the current. Often you will notice that some of these boats are swinging at anchor in different directions than other boats in the same harbor. This is because of the earth's rotation: **the tidal currents in such places may be roughly circular.**

If you could tether a one-mile-long string in the center of a typical harbor, and attach a floating buoy at the end, the buoy would not drift in and out of the harbor in a linear path during the tide cycle. Instead, it might drift in a continuous irregular circular path, such as the one in illustration 6-7. This is a common pattern, but it is not universal. Many other parameters affect the pattern of the current in each harbor: convection currents related to temperature gradients, salinity gradients related to freshwater run-off, and the shape and depth of each unique coastal basin. Because of the unique parameters of each harbor, the final pattern of the tidal current will not be perfectly linear nor perfectly circular, but the earth's rotation will tend to form circular currents within each coastal basin. These usually progress clockwise in the northern hemisphere, and in the opposite direction south of the equator.

(6-7)

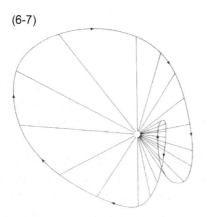

If you attach a float to a fixed point in the center of a typical harbor and allow it to move with the tidal current on a long line, it will move in a distorted circular path, such as this clockwise pattern of the actual current in San Francisco Bay.

There might be strange loops of current, or there could be localized areas where the flood tide current flowed out toward the mouth of a harbor, and the ebb tide current flowed in from the sea. Along the shoreline of coastal gulfs and bays, the current might be toward the north or toward the south during the incoming phase of the tide. This would depend on the unique circular pattern of tidal current in that specific basin, with all of its odd loops and reversals. These unique patterns would, in turn, depend on the specific configuration of the shoreline and bottom contour in that specific basin, which distort the circular current driven by the earth's rotation. **The direction of tidal current flow is somewhat unpredictable for each coastal basin because once the earth's rotation deflects their linear path to circular**

motion, the shoreline and the bottom contour further distort the circular flow into unique patterns with loops and reversals.

It is important to note that there are harbors and bays where the local currents from the constant outflow of rivers (and other local conditions) are stronger that the forces due to the earth's rotation. Such conditions may completely dominate these coastal currents, and produce unique local patterns that may or may not be circular. Just as whirlpools and tornados are local phenomena unrelated to the earth's rotation, not all rotating currents are due to the Coriolis force.

The tidal currents within rivers are linear, up and down the river, only because the riverbanks confine the water and prevent these currents from moving in a circular path.

Whereas the currents in coastal bays and rivers are derivative of the tides, the great ocean currents are not. The tide generating forces are too weak to affect the offshore ocean currents. The net lunar gravitational force (page 27) is only 1/11,700,000th of earth's gravitation at the ocean surface. As we shall see in the coming chapters, the small oceanic tidal range is enhanced as it passes over the continental shelf and moves onto the coast and into its gulfs, bays, and harbors. Only after the tidal energy is amplified by the geostrophic forces (forces due to the continents on a rotating globe) is it strong enough to cause the coastal currents.

Now we will turn to one final consequence arising from the rotation of the globe and the oscillation of the tides. **There is a transfer of energy between the revolving earth, the orbiting moon, and the moving tidal ellipse.** The dynamical effect is to slow the earth's spin (loss of energy), and increase the moon's angular momentum, by increasing the radial distance to the moon (gain in energy). This is about equal to the total power consumption of every country on earth. **This actually slows the rotation of the earth over eons of time**: by one second every 41,000 years. Research on the growth of ancient coral beds confirms that the earth did revolve faster, millions of years ago. On the primordial earth, a day probably lasted about six hours, and the moon was about 200,000 miles closer to the earth.

Resistance to the tidal movement over the surface of the earth can be harnessed to generate electricity, by the friction between the moving water and a turbine. This is limited to coastal basins with at least 16 feet of tidal range, and there must be a suitable location to dam off the entrance. Of course, there is controversy about possible adverse effects on the local marine biology.[c]

(6-8)

Tide powered electric generators on the Rance Estuary in France. In order to utilize the tidal power in this way, the range of the tide must be at least 16 ft., and there must be a suitable location to dam off entrance.

---

[c] An interesting theoretical consideration in the construction of tide-powered hydroelectric dams is that the friction from harnessing *all* of the tidal energy on the coasts of the continents would slow the rotation of the earth by a measurable amount.

On a more practical daily basis, this friction causes a global delay or lag in the position of the high water after the moon passes overhead. See illustration 6-9 below. In other words, **the high tide is not actually at the time the moon is overhead** — it lags behind a true alignment toward the moon by an average of three percent in the deep oceans. This effect is much greater over the continental shelves (about 8 hours at New York harbor, for instance). The tide tables account for this phenomenon, so that it is not of any practical significance to boaters using the tables. It is an interesting fact of academic significance, and is known as "establishment," or the "age" of the tide. There is a similar lag, or establishment, of the solar constituent of the tide.

While I was writing this chapter, I happened to be out on a dock, with a full moon directly overhead. I noticed that it was low tide. I had expected that the tide would be high at my location on the earth at times when the moon was overhead. However, because of "establishment," the tide is *never* high at the coast when the moon is directly overhead.

(6-9)

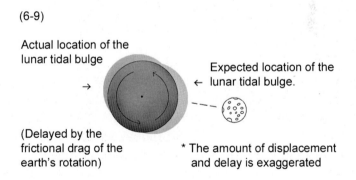

Actual location of the lunar tidal bulge →

Expected location of the ← lunar tidal bulge.

(Delayed by the frictional drag of the earth's rotation)

* The amount of displacement and delay is exaggerated

In our discussion of the amphidromes, we have referred to the term sea level. Sea level means different things to different people. If you are the operator of a pleasure boat along the coast, or a building contractor who needs to know how close to the waterline he can to build a house, you will use mean sea level (MSL). MSL is measured with respect to a benchmark on land. The location of the benchmark is chosen by the scientific community on a stable part of terra firma. The MSL is also the "still water level." This is the level of the sea with all of the high frequency motions (wind driven waves, ocean swells, and tides) averaged out.

This definition of sea level changes through epochs of time. The changes may be due to a real alteration in the height of the water (related to climatic temperature change, for example), or may be due to an alteration in the height of the land (due to tectonic shift, for example). Therefore, this value is reviewed annually to see if it requires revision; and by agreement among the scientific community, it must be actively considered for revision every 25 years. Most recently, MSL for North America was updated at the National Tide Datum Convention of 1980. Among other things, this convention replaced the National Tide Datum Epoch from 1941 through 1959 with the more recent epoch of 1960 through 1978. In addition, the convention lowered the chart datum from "mean low water" to "mean lower low water." MLLW is now the standard for the United States and adjacent waters.[d]

There is another way to access the sea level. If you are a geologist or an oceanographer, sea level means the local height of the water relative to a global reference surface *of the water* on the earth, called the "geoid." This global ocean surface, the geoid, is used instead of a benchmark on the land. The geoid is not a sphere for a number of reasons: (1) the earth itself is not a sphere.

---

[d] Mean low water (MLW) is the average of all of the low tide measurements for the past year. Recall that on a given day there may be a significant difference in the two low tides because of the declination of the lunar orbit (chapter 3). On these days, there is a higher low water and a lower low water. Mean lower low water (MLLW) is the average of the lowest low tide measurement on each day of the past year. This is a lower level than the previous MLW standand, and therefore safer for navigation over bottom hazards.

It has a greater radius at the equator than its radius at the poles. This is because centrifugal force of the earth's rotation is greatest at the equator, and because the earth is not entirely rigid. The bulge of the earth at the equator deforms the earth into an "oblate spheroid," and of course the water on the surface assumes this shape. (2) The gravitational field of the earth is not uniform. There are concentrations of mass in the earth's interior, and the surface of the earth is not flat, it has mountains and valleys. There is greater gravity on the water covering mid-ocean mountain ranges, and less gravity on the water over the ocean trenches. (3) The water is not motionless. The predominant trade winds and the ocean currents force the water to a higher level down wind and down current. (4) The circular motion of the tidal wave creates the amphidromes, with lower levels in the center. (5) Some oceanographers place importance on the gradient in the ocean's surface due to the tidal ellipse (the tidal wave, itself). Because the high tide may occur on one side of an ocean while the other side is at low tide, there is still another gradient in the surface. Movement of water down this slope may be one of the parameters that influences the formation of the amphidromes, mentioned earlier in this chapter. Tidal influences are averaged out of the global reference surface, the geoid; but it is still complicated, and it is not level because of the earth's bulge at the equator, the earth's gravitational inconsistency, and the trade winds and ocean currents.

All of this is used to establish sea level relative to the global reference surface of the water, the geoid. Instead of mean sea level (MSL), oceanographers refer to the National Geodetic Vertical Datum (NGVD). Landlubbers and coastal mariners should not be confused by the geoid and the NGVD; they should rely on the mean sea level or MSL, which is land based.

So, what is sea level? It is an average of about 16 inches lower in the Atlantic than the Pacific. It is about 8 inches higher on the west side of the Panama canal than on the east side. It is about 4 inches higher at Tampa, Florida, than at Daytona Beach, Florida, because the trade-winds that blew Christopher Columbus to the Caribbean force water into the Gulf of Mexico. It is about 36 inches higher at the northern end of the gulf stream than at the southern end, because this huge river in the ocean flows north and pushes the water higher down current (it must take energy from other natural forces to do this). **Sea level is a lot of things, but it is not level**. Remember that sea level is a constant value (although it must be revised every 25 years), whereas the range of the tide is a local fluctuation in the height of the water during each hour of every day.

The Coriolis Force and Oceanic Amphidromes

Supplement to chapter six:

The Coriolis force is not a real force: it is a fictitious force, or apparent force. Scientists make their observations on earth, while they are on the revolving, spherical, earthly frame of reference. Therefore, their direct observations may not be explained by Newton's Laws of Motion.

In order to simplify the mechanics of motion within the earthly frame of reference (a rotating sphere) physicists have taken the convention of adding two "apparent forces," centrifugal force and the Coriolis force. Centrifugal force is radial (outward), and applies to objects on the rotating earth, whether they are stationary (relative to the earth) or moving. The Coriolis force only applies to objects that are moving relative to the earth's surface. The Coriolis force is a deflection perpendicular (sideways) to the velocity of a moving object.

Because of the earth's rotation, these apparent forces must be taken into account along with the real forces acting on earthly objects. When you stand motionless on a scale, the force of the earth's gravitation acts on your body, but the effect of this force, your *weight*, also depends on the centrifugal force due to the earth's rotation — your weight is slightly different, at different latitudes (see page 29). If there were a perfectly straight (frictionless) highway from Texas to Canada, as you drove your vehicle at highway speed, both apparent forces, centrifugal and Coriolis, would be required to explain certain scientific observations: (1) the weight of your car would be influenced by centrifugal force, and (2) the path that your car traveled would be influenced by the Coriolis force. If you did not make steering corrections to counter this Coriolis force, your car would completely leave the highway toward the right after only a few miles. Since you are constantly making tiny steering corrections in order to stay in your lane, you are unaware that some of these corrections are necessary due to the Coriolis force.

The origin of the Coriolis force can be explained in terms of the conservation of angular momentum. For an object moving in a circle, the angular momentum (**L**) is the mass times the velocity times the distance from the axis (radius). Furthermore, the velocity is equal to the angular velocity **(ω)** times the radius **(r)**. Solving forward yields: the angular momentum is equal to the mass times the angular velocity times the radius squared, or **L = m ω r²**. This is an extension of the familiar law: momentum equals mass times velocity.

If you apply a force (**F**) to a mass in the earth's sphere of influence, it appears to an observer on earth as if the object is being acted on by an effective force (**Feff**) according to the formula:

Feff = F − 2m (ω × vr$_{ot}$) − (mω) × (ω × r).
where: ***m*** = mass of the object,
        ω = the (constant) angular velocity of the earth,
        vr$_{ot}$ = the velocity of the mass relative to the rotating set of axes,
        r = the radius from the origin of the rotating frame to the object.

The last term: **− (mω) × (ω × r)** is the centrifugal force. It is a vector outward and has a magnitude of **mω²r² sinθ**, where θ is the angle between the angular velocity vector **(ω)**, and the radius vector **(r)**. The middle term: **−2m (ω × vr$_{ot}$)** is the Coriolis force, which only applies when the object is moving relative to the earth. At first glance, it seems odd that this force would only affect moving objects. However, it makes perfect sense when you consider the following.

The earth's north–south polar axis is the axis of rotation of the equator and each latitude as well. Of course, since each latitude has a smaller circumference as you proceed from the equator toward the poles, each latitude has a smaller radius, proceeding toward the poles. As objects moving over the surface of the earth change latitudes, their angular velocity must increase as the radius decreases. An object moving away from the equator toward the north is also acted upon by a centripetal vector of force (gravitation) and is therefore constrained to follow the curvature of the earth. At its new latitude it is getting closer to the axis of rotation, and it increases angular velocity. This means that it is traveling faster than the ground under it, and it will move over the earth's surface toward the east, and curve to the right. By the same logic, it will decrease angular velocity as it travels toward the equator (south) and, going slower than the surface, it will move toward the west — again, curve to the right.

But what about motion toward the east and west? This is only slightly more complicated. Moving toward the east, the velocity of the object adds to the rotational velocity due to the turning of the earth. In the reference frame of the earth, this creates more centrifugal force, which forces it out radially. Since there is the additional centripetal vector of force due to the earth's gravitation and it cannot move outward, it turns toward the equator where it is further from the axis — again, turning to the right. Westward motion causes a loss of centrifugal force, and it turns away from the equator, to a smaller radius — toward the right. For those readers who wish to pursue a more thorough and mathematical treatment of this phenomenon, the bibliography for this chapter includes numerous excellent resources. See: Feynman, Richard P.; Goldstein, Herbert; Marion, Jerry B.; Sommerfeld, Arnold; Stommel, Henry; Swartz, Clifford E.; and Synge, John L.

Chapter Seven

The Seiche Effect and Basins of Oscillation. Tidal Intermixing

"The sun's a thief, and with his great attraction robs the vast sea; the moon's a thief, and her pale fire she snatches from the sun; the sea's a thief, whose liquid surge resolves the moon into salt tears."   William Shakespeare

One of the most curious tidal patterns is the occurrence of one high tide each 24 hours that is found in the Gulf of Mexico, as well as some basins on the coast of Southeast Asia. This is explained by the "seiche effect," which occurs within "basins of oscillation." The seiche effect is commonly known as the bathtub effect. When you climb out of the tub, for a brief time the water is lower on the end that you exit. Before the water levels itself, it rocks back and forth in a series of waves, an oscillation known to oceanographers as a seiche (pronounced: sigh-sh).

When the barometric pressure changes suddenly at one end of a large lake, a sort of tidal motion is induced in the surface. In mountainous Switzerland, there are very sudden atmospheric pressure changes that tend to have a dramatic effect on Lake Geneva, which is large and elongated. The residents of Lake Geneva coined the word "seiche" (meaning dry) to describe these "tides" that left the shores of their lake. Much of the original work on The Dynamic Theory of the Tides was based on the observations of Laplace at Lake Geneva. The seiche effect was relatively easy to isolate and understand when it was caused by changing weather over a lake. It soon became apparent that this phenomenon was one of the reasons for the variety of the tides in the oceans, seas, gulfs, and all other "basins of oscillation" around the globe.

When I first learned of the concept of basins of oscillation, I immediately thought about my experience with surf fishing. When fishing on the beach, it is common to first net some live baitfish. These must be kept in a bucket and transported down the beach to your chosen fishing spot. If you put a bucket of water in your automobile trunk and drive along the beach, the water will invariably begin to oscillate (slosh out into your car). If you use a tall deep container, the oscillation of the water will be different than the result with a wide shallow washtub. Each ocean, gulf, or bay is a basin of oscillation, and each is different in length (east to west), width (north to south), and depth. It is not surprising that **in each basin of oscillation around the globe, the seiche effect has a different pattern once the water is set in motion by the rhythmic celestial cycles**.

In many ways, the tidal waves arriving at the continental coast are similar to the ocean swells that slow down and gain height in shallow water, forming the breakers at the beach. However, in one important aspect, the tidal waves are very different. When the breakers land on the soft sandy beach, the wave energy is completely dissipated. By contrast, **the tidal waves encounter the solid continental shelf and they bring far too much energy to be absorbed at the coast. Therefore, these waves are reflected back toward the next tidal wave.**

One of the most important principles in our understanding of the tides is that the tidal motion of the water in each basin around the globe never has time to settle back into a resting equilibrium before the next wave of the tide arrives at that point on earth. Therefore, in each basin of oscillation, waves resulting from the previous tide may interact with the waves of the following tide. This is more likely in smaller basins, which are confined by a landmass on many sides, such as the Gulf of

Mexico. Thus, the interference of seiche effect waves oscillating back toward the east may *cancel out* waves of the following tide, moving toward the west. This can result in only one high tide each day. Later in this chapter, we will look at the details of this phenomenon in the Gulf of Mexico, but first we must deal with some basics.

It is of critical importance in deciphering such complex tide patterns to understand that **there is not just one oceanic wave of tidal energy circling the globe from east to west. In fact, there are multiple independent tidal waves, each forced by its own celestial cycle:** the daily rotation of the earth, the moon's elliptical orbit, the cyclical change in lunar declination to the equator, the earth's elliptical solar orbit, etc. **Each of these waves of tidal energy, each underlying constituent of the moving tidal ellipse, is physically there. They are not mere theoretical or mathematical concepts**.

It is difficult to think of multiple, independent, simultaneous waves of energy circling the globe, each driven by a different cycle of the solar system. Our intuition prefers one tidal wave each twelve hours — this correlates with the tide tables, and matches the flood and ebb at the shore. However, it is absolutely vital to our understanding of complex tide patterns to deal with this fact: **as it changes minute-by-minute, the height of the water is the sum total of multiple independent, simultaneous waves of energy moving through the water, each with its own frequency and amplitude, each making a contribution to the measured water level.**

In order to convince ourselves that these are real physical waves, rather than mere mathematical theory, let's consider the very real phenomenon of "rogue waves." Many seaworthy ships have been swamped by rogue waves. Because they are completely unpredictable, and may be impossible to negotiate, a captain must depend on all of his skill and lots of luck to survive these watery monsters. Rogue waves may occur whenever two or more wave patterns are moving over the same surface, at the same time. For instance, suppose that a small open powerboat is cruising offshore. There are gentle three foot swells, slowly rolling in from a storm that occurred last night, 100 miles beyond the horizon. In addition, there are two to three foot wind-driven waves coming from offshore. These are irritating, but pose no hazard. About a half-mile away, a large fully loaded cargo vessel plies along, outside of their position, parallel to the coast. The wake of this ship is only about two feet high by the time it reaches our cruising watermen.

The odds are against it, but by pure chance, the crest of the swell, the crest of the wind-driven wave, and the crest of the wake all pass through the exact same point on the surface at the same instant. The three waves reinforce each other and create a rogue wave six feet high. Later the captain would testify, "This one monster wave rose up all at once, and rolled us over." One monster wave, indeed. Try and convince that captain otherwise — forget about the reinforcement of three separate constituents — that *was* "one monster wave." Likewise, **we see the tide rise and fall in one continuous waveform of flood and ebb. But, beneath the surface, the constituent waves are each making their contribution.**

You may protest that you can *see* the three surface waves in the above parable. Well, you can actually see undulations on the surface due to some tidal constituents after they reach shallow water. I have seen them several times.

The oceanic tidal gravity waves have very long frequencies, because they are derived from the hourly, daily, monthly and yearly cycles of the near solar system. These ocean waves are not visible to us; the following crest is usually beyond the horizon. However, tidal waves in shallow water have shorter frequencies. This is explained in chapter eight. These shallow water waves may then create additional waves at still shorter frequencies, at the mouth of still shallower basins. It is the occasional surface undulation at these shorter frequencies that may allow us to actually see some constituents of the tide, on windless days, in flat shallow water.

These are *visible* frequencies of tidal energy manifest as surface waves. If you are out on the water on a regular basis, I am certain that you have seen them yourself. Doubtless, you assumed that all waves are either offshore swells, wind-driven waves, or wakes. Now that you know otherwise, if you will look for tidal energy waves on windless days, in flat shallow water, you will recognize them. They are usually rather subtle, although they may be dramatic, as shown below.

(7-1)

High frequency waves moving up a river in France. This allows you to see one of the constituents of the summation tidal wave.

Since you probably haven't recognized this evidence of the multiple tidal energy waves underlying each tide pattern, I will offer the fact that convinced me of their physical presence. There are many strange and unlikely tidal patterns around the globe. One of them is so bizarre that the local oceanographers, in Thailand, felt compelled to find an explanation, and report it in the scientific literature. The Gulf of Thailand is a basin on the margin of the China Sea. As in all basins, there are tidal energy waves of diurnal frequency and waves of semidiurnal frequency progressing through this gulf every day. However, in this gulf the semidiurnal waves progress in a counterclockwise direction, and the diurnal waves progress in a clockwise direction. That is unusual. As a rule, the earth's rotation creates counterclockwise rotation of the tidal waves in the northern hemisphere. It is not important to understand the complex local hydraulics that direct the diurnal tidal wave clockwise in this place, but the fact that these two simultaneous waves rotate in opposite directions is very instructive. Because this separates the two energy waves, it allows us to unequivocally see them as physically present, independent, simultaneous, and real.

In spite of all this, you may still argue that it is relatively easy to accept how lunar gravitation and solar gravitation each create separate and distinct tidal waves, but it remains less obvious that the complexities of the solar system, like the declination of the lunar orbit and the earth's elliptical solar orbit, each create their own independent, physical tidal waves.

To begin with, we can say for certain that the shape of the global tidal wave would be different if you excluded the contribution of any of these celestial constituents. That being the case, consider the following analogy. Suppose that you are looking at a beam of green light. Then, someone uses a filter and removes the yellow light wavelength from the beam. Of course, now you will see a blue light coming out of the filter, proving that the green light is actually composed of yellow wavelength and blue wavelength light.

Using a prism, we can demonstrate that white light is composed of a rainbow of independent light waves with red, yellow, blue, and violet wavelengths. Looking in a physics textbook, in the chapter on optics, we find a list of the wavelengths of red, yellow, blue, and violet light. But, we were interested in the white light, the light that we saw pass into the prism, and we find that the white light wavelength is not on the list. We ask the physics professor, "What is the wavelength of white light ?" He gives us an odd look, and says, "There is no white light wavelength. White light is just a *perception*, the way you interpret the condition when all of the different color wavelengths are present at the same time."

Our first thought was that the tidal wave, the one that lifts our boat, the one that fills up our harbor and is depicted in our tide table, that wave is physically real; and the subtle constituents arising from gravitation and momentum seem to be mathematical concepts. Now it is clear that *the opposite is true*. **It is the waves caused by these subtle constituents that truly move the oceans; and when they are all present, and all shape the oceans at the same time, together, they create the *perception* of a single global tidal wave.**

We are now ready to explain the unusual diurnal tide in the Gulf of Mexico. The seiche effect in the Gulf of Mexico can create a diurnal tide only because there is a real physical *semidiurnal* wave of tidal energy (driven by a lunar celestial cycle), and another real physical *diurnal* wave of tidal energy (driven by a solar celestial cycle), both entering the gulf together toward the west.

(7-2)

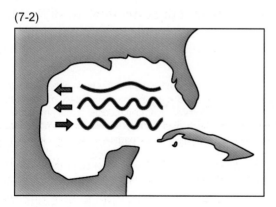

Diurnal tidal wave (red) and semidiurnal tidal wave (blue) within the Gulf of Mexico.

The diurnal wave is then reflected from the western shore and travels back toward the east. The *period* (time for one complete cycle back and forth) of this diurnal seiche in this particular basin is about 24 hours — waves slow down in smaller, shallower basins. Of course, 24 hours is also the period of the global diurnal tidal wave. Just as this reflected seiche returns to the eastern entrance into the gulf, the following diurnal tidal wave arrives there. Because the internal seiche and the external forcing diurnal wave have similar periods, a diurnal pattern is supported in this gulf. This hydraulic process will be examined in greater detail in the following chapter.

The semidiurnal wave in the Gulf of Mexico also generates a seiche moving back toward the east. By pure chance, this reflected wave is perfectly out of phase with the semidiurnal wave of the following tide into the gulf. See illustration 7-2. Therefore it cancels out the next incoming semidiurnal wave (recall aphasic wave interference in chapter five). This process leaves only the diurnal tidal wave pattern, and one high tide a day in the Gulf of Mexico.[a] Thus, **diurnal tides and many other unusual tide patterns may result from reflected seiche effect waves traveling back against the flood of the following tide cycle, causing interference that cancels out some tidal constituents, leaving other constituents to create a unique tide pattern.**

---

[a]Interference of aphasic waves may be an oversimplification of the hydrodynamics within the Gulf of Mexico. The oceanography literature contains more complex mathematical treatments of the tide pattern in this basin. However, hydraulic models of this phenomenon generally depend on the following: In any such basin, there will be a natural period of oscillation, with a proscribed amount of time required to propagate a wave back and forth across the basin. This natural rhythm will largely depend on the length and depth of the basin. Then, the internal waves arising from the basin's natural period of oscillation (the seiche) must interact with the external forcing tidal waves, which have their own period and frequency. Sometimes these internal and external waves have periods that are attuned or synchronized with each other, and sometimes not. This will be discussed in greater detail on pages 135–136. In the Gulf of Mexico, the diurnal seiche is synchronized with the diurnal period of the ocean tide, supporting the diurnal pattern, while the semidiurnal seiche is "out of synch" with the global semidiurnal tide, and the semidiurnal pattern disappears within the Gulf of Mexico.

It seems intuitive that waves moving in the opposite direction might interfere, and cancel each other out. However, it is also possible for advancing waves and reflected waves to *reinforce* each other, if the *crests* of the two waves meet near the same point on the surface. In fact, this is a very common phenomenon within the waters between the continental shelf and the coast (shelfwaters). If you decide to look into the more advanced literature and textbooks on the tides, you will find repeated reference to the fact that, "The tide pattern within the shelfwaters and within gulfs and coastal basins can be considered as a standing wave with a superimposed progressive wave." This is absolutely accurate, but this sort of encoded scientific jargon is meaningless to anyone but an oceanographer. Let's try to translate the statement, "a standing wave with a superimposed progressive wave" into English.

Consider that the reflected tidal wave is almost a mirror image of the incoming tidal wave. It is *not* identical with the incoming tidal wave pattern because it has less energy due to friction and damping that occurs along the coast. This damping includes relatively small forces such as absorption of energy by the somewhat elastic planet, deflection of some of the energy toward other directions by the earth's rotation, and the friction within very shallow estuaries. Furthermore, some of the energy of the reflected wave is lost back into the ocean.

When two *identical* waves are moving in opposite directions, this could have either of two possible results. They might cancel each other out by interference, if the crest of one wave met the trough of the other wave; or they might reinforce each other into a larger wave, if both crests meet near the same point over the bottom. See illustration 7-3(a), below. In this case they would create a *stationary* waveform, an elevation in the surface of the water at the point where they meet. In other words, when opposing waves are moving in opposite directions, this does not necessarily mean that they will interfere, they may reinforce each other; but in either case, the resulting wave will be stationary. This stationary heightened water level is called a "standing wave."[b]

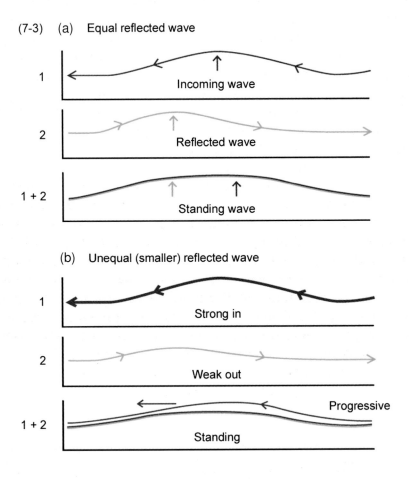

(7-3) (a) Equal reflected wave

(b) Unequal (smaller) reflected wave

[b] We have discussed standing waves due to the crest of an incoming wave meeting the crest of a reflected wave. Mariners may have experience with standing waves due to an entirely different cause: when a *strong* current is flowing over a *very abrupt* change in depth of the bottom, a standing wave may form on the surface.

But, what if the advancing wave and the reflected wave are not identical? See illustration 7-3(b). If one of the waves is stronger than the other (like a flood tidal wave, that is stronger than reflected wave going back in the opposite direction), then a standing wave may still occur where they meet (over the continental shelf, for example). This would elevate the water level over the entire shelf or basin. However, there will be some energy left over from the stronger wave (the flood tide), and this energy will progress in the direction that the stronger wave was traveling. This "leftover energy" progressive wave will not have much height of its own, but it will be superimposed on the elevated water level of the standing wave, that has the energy of both the flood tide plus the reflected wave from the previous tide.

Therefore, the statement, "a standing wave with a superimposed progressive wave" means that **the water level over the continental shelf may be elevated by a standing wave, created by combining the advancing flood tide and the reflected wave from the previous flood, while the moving cycle of the tide is preserved by a progressive wave moving through this elevated shelf water..** This is a very important phenomenon. It is one of the several mechanisms by which the tidal range is amplified over the continental shelf and within coastal basins. It is also another example of the principle that the energy of successive tidal waves may mix with each other, and create a variety of patterns along the coast.[c]

---

[c] Oceanographers commonly refer to "a standing wave with a superimposed progressive wave" to explain the tidal elevation and the tide pattern over the continental shelf. In this context, "progressive wave" means the movement of tidal energy through the continental shelf waters. It does not mean that there is horizontal movement of the water itself. Unfortunately, there is another use of the term "progressive wave" in hydraulics. In the classification of waves like ocean swells, wind driven waves, and the like, we have referred to "oscillatory waves" in which energy moves through standing water (a wave moving under a pelican on the surface), and "progressive waves" in which the water moves horizontally over the bottom (the breakers and the water moving up onto the beach at the waterline). In this context a progressive wave involves moving water, rather than energy moving through water. This is the common usage of the term "progressive wave."

The influence of the seiche effect within basins of oscillation is not confined to gulfs, bays, and continental shelfwaters. These forces are very important in the oceans as well. **The major reason that the coasts of the Atlantic and Pacific oceans have such different tide patterns (diurnal, semidiural, mixed) is that each ocean basin selectively responds to different frequency tidal waves.**

We have discussed several phenomena that contribute to the way in which different celestial cycles dominate the tide pattern in different oceans. (1) Wherever tides occur, the motion of the water does not have time to come back to a resting equilibrium before the energy of the following tidal wave arrives. Therefore, reflected waves are oscillating back toward the east and mixing with tidal waves moving toward the west in each basin. (2) The Atlantic and Pacific oceans are very different in length and depth. (3) The tidal wavelengths are on the same scale as the size of the oceans, being in the range of 10,000 miles from crest to crest. (4) The diurnal tides have wavelengths that are twice as long as the semidiurnal tidal waves. (5) The diurnal tidal waves are often related to solar influences, and the semidiurnal waves are generally due to lunar cycles.

The Pacific Ocean basin is much larger than the Atlantic basin. It is logical that the natural period of oscillation of **the larger Pacific basin selectively responds to the longer wavelength of the diurnal tide, whereas the smaller Atlantic basin is more attuned to the shorter semidiurnal wavelength.** Recall that the tidal wave is traveling at 430 miles per hour. The semidiurnal period (between successive waves) is about 12 hours. Therefore, this wave will travel about 5,000 miles in 12 hours — this is approximately the distance across the Atlantic Ocean. The diurnal period is 24 hours, and during this time this wave will travel about 10,000 miles, which is in the range of the distance across the Pacific Ocean. We have just learned that the tidal wave does not travel straight across these oceans, but rather, there are roughly circular patterns of oscillation around the global basins. Therefore, these numbers do not strictly link these wavelengths to these oceans, but they do show proportionality between them.

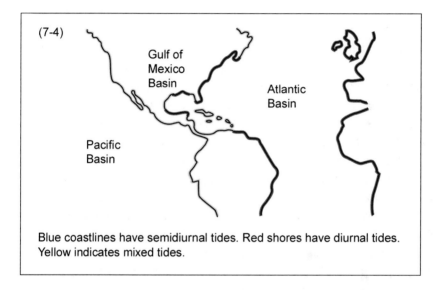

(7-4)

Blue coastlines have semidiurnal tides. Red shores have diurnal tides. Yellow indicates mixed tides.

As it turns out, most of the tides *are* semidiurnal on the Atlantic coasts. However, most of the Pacific coasts have *mixed* tides, *not* diurnal tides. See illustration 7-4, above. To understand this, recall that the lunar gravitation at the surface of the earth is twice as strong as the solar gravitation. The Atlantic basin is closely attuned to the shorter lunar semidiurnal cycles, *and* the lunar gravitation is strongest. Therefore the Atlantic coast tide patterns are dominated by one group of celestial cycles, lunar. This keeps the Atlantic coast tidal patterns relatively simple. On the other hand, in the Pacific Ocean, the longer diurnal rhythms are attuned to the larger deeper basin, while the lunar semidiurnal rhythms are preserved in the Pacific by the stronger lunar gravitation. The result is that neither solar nor lunar cycles will dominate.

Therefore, the Pacific coast tide patterns are mixed. They are more complex and less consistent from day to day, because there is constant reinforcement and interference of two strong rhythms.

Because each basin of oscillation is attuned to different celestial cycles, some astronomical influences dominate the tide tables at some harbors, whereas other astronomical influences dominate the tide tables at other harbors, if these harbors are in different basins of oscillation. To further complicate this tidal influence, the mountain ranges on the ocean floor in both the Atlantic and Pacific Oceans divide these oceans into multiple basins. This contributes to the complex variation in tide patterns on the coasts of the continents around the globe.

The tides at mid-oceanic islands are unique. Let's look at the tides on two mid-ocean islands, Tahiti in the Pacific, and Bermuda in the Atlantic. On the shores of Tahiti the tides have a very small range. The Tahiti tides are diurnal and repeat almost exactly the same pattern every day. If you drive a stake in the sand at the waterline in Tahiti, you could mark the stake with lines as the flood tide came up, and use these marks (the height of the water) *to tell the time of day* for a long time afterwards. This is because the Pacific tides at Tahiti are entirely dominated by the diurnal constituent called K1, which has a period of 24 hours. This occurs in this place because the size and depth of the Pacific basin allows the longest wavelengths to dominate the tide pattern. The small tidal range occurs because of the close proximity of the nearest amphidrome to this island, resulting in a minimal tidal range near the center.

In Bermuda the tides are semidiurnal, and have significant daily and monthly variation. The Atlantic basin is attuned to the period of the semidiurnal lunar constituent called M2. The Bermuda tidal variation results from the same lunar cycles of declination to the equator, the elliptical orbit of the moon, and other lunar celestial cycles that effect the entire Atlantic basin. The height of the tidal range in Bermuda is atypical for a midocean island — it can be as great as three feet. One of the reasons for this large range is that Bermuda is located at the outer periphery of the nearest amphidrome in the north Atlantic.

You may never have any reason to know the tide patterns at these remote islands. However, they are wonderful examples of how every tide results from a combination of the fundamental forces that we are accumulating in these chapters. In order to grasp these concepts we sort of trivialize them as "the bathtub effect" in different basins that are attuned to different celestial cycles, and great "whirlpools" in the oceans that are higher at the outer edges. But these forces are very real, and they are part of the explanation for the variety in the real tidal patterns seen around the globe.

In earlier chapters we conceptualized the tide as a great wave of energy moving westward around the globe. Now we understand that **the tidal energy is: (1) composed of multiple wave frequencies, (2) turned in a circular path by the earth's rotation, (3) deflected by the continents, (4) oscillating back and forth within a basin, and (5) interacting with the tidal energy of the previous tide. In the following chapters we will learn how (6) the height of the water is amplified by hydraulic forces over the continental shelf, and (7) the pattern of the tide cycle may be distorted by the friction within shallow estuaries.**

Chapter Eight

Coastal Geography and Near Shore Topography, Resonant Co-Oscillation, Sustained Forcing

"Foolish consistency is the ...goblin of a small mind."

Ralph Waldo Emerson

At the end of the nineteenth century, classical Newtonian physics was being supplanted by the new subatomic physics. This was an exciting time, when new discoveries were reported every year in the scientific literature shared by physicists around the world. Most of the progress came from Europe, but two of the leaders in this quest to unravel the mysteries of the universe were Albert Michelson and Edward Morley, who worked together in America.

The physical world consists of matter and energy. Matter is constructed from atoms, and many of the best young minds wrestled with the new subatomic particles. Michelson and Morley had tackled the problem of energy, specifically how energy travels through space. They were certain that light travels from the stars to the earth as waves, just as energy travels across a pond in waves when you toss a stone in the water and the ripples reach the edge. There was one big problem, however. If light travels through empty outer space in waves, the next question was, "waves of what?" Lacking any experience with outer space, scientists had filled the void with a theoretical substance called "the ether." This theoretical ether filled the empty space between the stars and planets, and was the stuff that the waves travel through — there were waves of ether.

It was already known that sound waves in air will change pitch or frequency if the source of the sound is moving. This is the Doppler effect that makes a train whistle change pitch as the train passes by you. Michelson and Morley decided to measure the change in the speed of light that was traveling from different directions on the moving earth. They took two large heavy instruments to the top of two mountains, and measured light going from one to the other. Then they had to carry their instruments up other mountains in order to measure light going in other directions.

Their theory was that since the earth itself was speeding through the universe within our galaxy, once they aligned the light in the direction of the earth's motion, the speed of their light would change. This was a brilliant experiment, because it would prove the wave theory of light and confirm the existence of the ether.

However, when they sent their light in different directions, it did not change at all. And so, they hauled their equipment up some more mountains, and measured the light going in still more directions. But no matter how many damned mountains they dragged up the damned instruments, the light never changed. All they could do was report the results of this experiment in the world literature: a report with no real conclusions.

Young Albert Einstein had some training in mathematics and physics, but not enough to be invited to work at a university. He was working as a clerk in a patent office, and he read the report by Michelson and Morley. Of course, every physicist at every university in the world had also read this report of what they all considered a failed experiment. Young Einstein thought otherwise. He thought awhile about what Michelson and Morley had done, and published a paper that explained their findings and started the greatest revolution in science since Isaac Newton. In so many words, he said the following. The Michelson–Morley experiment was a success, and their results were accurate, but they were *asking the wrong question.*

He reasoned that *if* they had been trying to prove that the speed of light was a universal constant, which was immutable and unaffected by any other parameter or force, this experiment would be a great success. In fact, they had produced a brilliant proof of a monumental new fact: the universal constant of the speed of light. Einstein was an unknown and unproven upstart, but the irresistible logic of this idea was enough to attract the attention of the scientific world, a world that Einstein would soon change forever.

Scientific inquiry is just that: inquiry. In order to make scientific progress, the most important thing is to ask the right question. **In order to understand why the tides are so different at different times and different places, we must ask this question: why are the tides higher at the coast than in the deep oceans?**

In chapter one, we introduced the remarkable fact that the gravitation of the sun and moon together are only capable of causing about 18 inches of vertical displacement of the water on the surface of the earth. This is the measured range of the tide in the middle of the ocean. All very interesting, but somewhat academic. A mariner on the ocean doesn't really need to know if the depth is 5000 feet or 5002 feet. No one buys a tide table for the mid-oceanic longitudes. The tide tables are for the coastal waters, because (1) mariners need to avoid being grounded there, and (2) there is enormous variation in the pattern of the tides, the time of their arrival, and the height and range of the tides at each point on the coast of each continent.

Recall our analogy between the coastal tides and musical instruments. We said that, "The energy of these (musical) sound waves is taken from the breath and action of the musicians, just as gravitation sets the oceans in motion. But when the breath enters a flute, the music is very different from a trumpet. When the motion is exerted on a violin, the sound waves are very different from a cello. And, when the great energetic wave of the oceanic tide reaches the coastline of a continent, with its myriad beaches and bays, the tide is played out in myriad patterns."

We also said that, "These different (sound wave) patterns are all energized by moving air, which is essentially the same as it *enters* an oboe or a clarinet. However, the acoustic waves *within* each instrument are unique, because the instruments are different lengths and shapes, and made of different materials, just as the Mediterranean Sea and the Bay of Fundy have unique geography and hydraulics."

Before we consider the geographic features of your homeport that shape the local tide pattern, we will first need to learn what happens to waves when they enter shallow water. The laws of hydraulics define shallow water as a depth of less than 300 feet. At this depth, waves are constrained by different physical laws than in deep water. As the waves of tidal energy approach the coast, the shallow water laws of hydraulics change the character these tidal waves, just as they change the familiar ocean swells and wind driven waves that we see at the beach.

(8-1)

Imagine that you are at the salty shore, enjoying the mild breeze, listening to the hypnotic sound of the surf, and staring aimlessly at the breakers. Now hold a seashell up to your ear. You may use a coffee cup if you are in the city (do not let your family or your dog see you do this). The waves come from the horizon in an unbroken ceaseless pattern. Offshore the surface is only slightly undulating. Then, at a vague distance offshore, the crests and troughs of each wave become clearly defined, as they line up parallel to the shore. Although there is a relentless progress toward the land, each wave seems to slow down; the waves are crowded together as each wave swells up higher, finally cresting and falling on the beach.

The global waves of the tide cannot be seen because they are thousands of miles from crest to crest. The energy of these waves is literally astronomical. They are approaching the coast at about 430 miles per hour.[a] Everything about these titanic waves is on a grand scale. They begin to interact with the coastline far out at the continental shelf. However, they are governed by the same laws of hydraulics as the ocean swells and the wind driven waves that: (1) bend parallel to the beach, (2) slow down, (3) crowd together, and (4) swell up higher as they enter shallow water.

It must first be stated that hydraulics is very complex science that is not for the faint of heart. The language of hydraulics is full of calculus, sines, cosines, and other arcane mathematics. A thorough scientific explanation of wave theory is beyond the scope of this book. It has been said that scholarship is the art of

---

[a]When you encounter a wave of tidal energy in deep water, it is an "oscillatory wave," rather than a "progressive wave" like the breakers on the beach. There is no net horizontal motion over the seafloor in oscillatory waves. If a pelican is sitting on the surface, it does not move forward with the oscillatory wave; it merely bobs up and down. This is a wave of energy, transferred through standing water. The tidal waves move across the oceans at 430 miles per hour, without advancing the water beneath the ships at sea. Only in very shallow water, like the breakers and the water line at the beach, do the waves become "progressive" and move water forward over the bottom.

exhausting your subject without first exhausting your readers. In this chapter more than others, we must be satisfied with "getting a feel" for the matter at hand. In spite of these difficulties, we must attempt to gain at least a superficial understanding of this essential question: how and why do the deep ocean waves change when they move into shallow water?

To begin with, as a wave approaches the coast at an angle to the shoreline, it is refracted (bent) so as to be parallel to the shore. This is because the end of the wave nearest shore is in shallower water and slows down first. The end of the wave farthest from shore is now going faster, and it pivots around the slower shallow end of the wave. This rotates the wave to a parallel alignment. Once both ends of the wave are the same distance from the land, both ends move at the same speed from then on. Now the wave volume and wave energy are beginning to be concentrated near the shore.

As a wave enters a shallow depth, it begins to be reshaped by the bottom. Strictly speaking this is not actually due to friction, since none of the wave energy is imparted to the bottom.[b] Beneath the surface undulation that we associate with waves, there is a less obvious wave component known as "surge." Beneath the crest, the surge energy moves in a circular path. Actually there are a series of circles, extending down from the crest toward the bottom (see illustration 8-2).

---

[b] It may be somewhat surprising that a wave system can be altered by a shallow depth without some energy being expended as bottom friction. However, we are accustomed to other examples of how physical laws of nature dictate the shape of things. Examples: In outer space, a drop of liquid always assumes a perfectly spherical shape. A (magnified) snowflake always has a geometric shape composed of straight lines and angles, never curved lines. A fast moving boat applies steady constant pressure against the water, but the deformity of the water (wake) assumes the shape of a series of undulations. The wind blowing across the water or the desert does not pulsate, it blows steadily; and yet, the surface of the water or sand takes the shape of a series of waves or ripples. In a similar fashion, as the depth shallows near the shore, the circular movement of the surge near the bottom is distorted, and assumes an oval or ellipsoid shape. This begins a complex hydraulic process that creates a slower, higher wave crest.

As the depth becomes shallow, these circular movements are distorted, although no energy is lost. They assume an oval or ellipsoid shape, and this begins a complex series of events that leads to a slower but higher wave. The net result is to convert kinetic energy (less forward motion over the bottom) into potential energy (higher wave). **Remember, waves not only *have* energy; it could be said that waves *are* energy. They are a combination of potential energy and kinetic energy.** The potential energy arises from the height of the wave, just as a water balloon has more potential energy as you take it higher up to the third floor window. The kinetic energy arises from the speed of the wave, just as a faster motorcycle can jump farther over a canyon.

(8-2)   Particle motion in a wave

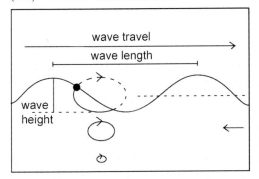

Circular motion of water between the crest and the bottom = "surge."

Now, as a wave of tidal energy arrives from the ocean, and enters the shallower coastal zone, it slows down and has less kinetic energy. Since it does not impart any of its energy to the bottom (another way of saying there is no friction), it retains the same total amount of energy. **The only way that a wave can conserve its total energy while it loses kinetic energy (slows down) is by increasing its potential energy. Increased potential energy means nothing more or less than a higher wave.** All of this concentrates the volume and raises the water level in the coastal zone.

In order to fully appreciate the effect of this hydraulic principle, we need a more dramatic example than the breakers at the beach. The wave caused by the greatest volcanic explosion in modern times, at Krakatoa, would have traveled across the deep oceans for thousands of miles. As this wave rushed over depths of 10,000 feet, it would be moving almost 500 miles per hour, but it would be only a few feet high, and the slope of the surface would be so gentle that it would pass unnoticed beneath the ships at sea.

However, Krakatoa exploded off Java, in the Sundra Strait, which averages 500 feet deep. The resulting tsunami started out about 10 feet high and traveling about 100 miles per hour. When it arrived at the 20 foot depth on the coast of Sumatra, it had slowed to about 20 miles per hour and formed a "giant black wall of water," according to survivors up in the hills. This wave was scientifically confirmed to be 115 feet high. It picked up a small ship with 28 crewman and deposited it in the jungle, one and a half miles from the coast, at an elevation 60 feet above sea level, on the day that it killed 35,000 people.

Our tidal wave, the ellipse in the earth's oceans generated by the powerful forces that control the near solar system, is likewise about 18 inches high and moving at 430 miles per hour toward the continents. It too must lose kinetic energy (speed) as it moves over the coastal shallows, and it too gains potential energy (height) within our ports and harbors.

Next the wavelength (distance from crest to crest) decreases. This is because **when the advance wave slows down, the following wave is going faster and catches up**. Now we need to think of the energy of the entire "wave system." By this we mean that the wave energy is actually measured throughout one entire wavelength — the energy within one crest *and* one trough, not just the energy beneath the crest. Again, there must be conservation of energy, and so **when the wave height increases, the wavelength decreases, crowding the waves together and concentrating the tidal energy near the coast.**

Finally, for completeness sake, we note that the "period" (time between passage of successive crests over a point on the bottom) does not change. This follows from the fact that the crests are moving slower, but they are closer together — a slower runner will cover a mile in the same time period that a faster runner does a mile and a quarter.

We can summarize all of the above shallow water laws of hydraulics as follows. **There are physical laws that reshape all waves as they enter a shallow depth. The overall effect is to slow the waves and increase their height. In the same manner, the shoaling of tidal waves near the coast amplifies the range of the tide relative to the deep ocean.**

Once the tidal wave passes over the continental shelf and begins to interact with a landmass, there are several local influences on the range of the tide. We will deal first with the two simplest and the most common geographic parameters. **The most common explanations for a difference in the range of the tide at different ports and harbors are: (1) the contour of the coastline — a straight coastline oriented north and south, versus a funnel-shaped coastline receding inland, and (2) the near shore topography — the slope of the ocean bottom near the shore.**

The contour of the coastline (straight or curved) has a major influence on the height of the tide. See illustration 8-3. The relatively small oceanic tidal range is not changed by the fact that it approaches a straight coastline. However, **when the advancing oceanic tide enters a coastline which recedes inland and progressively narrows into a funnel shape, a large**

amount of tidal energy is compressed into a diminishing space, and this greatly amplifies the tide.

(8-3)   Straight coastline   Receding coastline

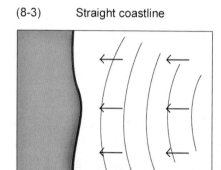 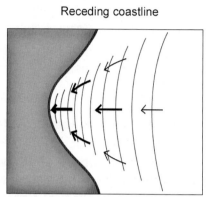

An excellent example is the Bristol Channel, England, shown below. This coastal basin progressively narrows as it extends farther and farther inland from the ocean. This results in one of the world's highest tides, with a range of over thirty feet at Bristol.

(8-4)

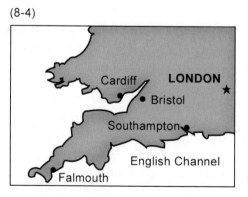

This funnel shaped channel has one of the most extreme tides on earth.

These tides range up to 35 feet at the narrow "head" of this funnel shaped waterway — above Bristol.

The highest tidal range on earth is at the head of the Bay of Fundy, shown below.

(8-5)

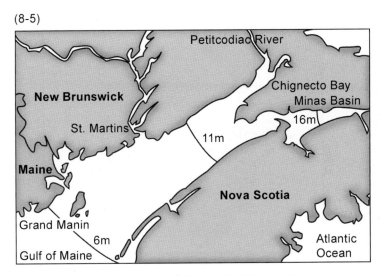

The Bay of Fundy is an arm of the Gulf of Maine. The average tidal ranges are shown in meters.

Note the progressive amplification of the tidal range from 6 meters (18 feet) near Grand Manin Island, at the mouth, to 11 meters (33 feet) at St. Martins, halfway up the bay, to over 16 meters (50 feet) at the head of the Minas Basin.

One of the major reasons for the Bay of Fundy's enormous tidal range of 45 to 50 feet is the funnel shape of this coastal basin which narrows inland from a wide mouth at the Gulf of Maine.

The contour of the coastline has a very significant effect on the range of the tide, even if it is not such a dramatic funnel shape. The coastline between Cape Hattaras, North Carolina, and Miami, Florida, recedes gradually toward the west.

(8-6)

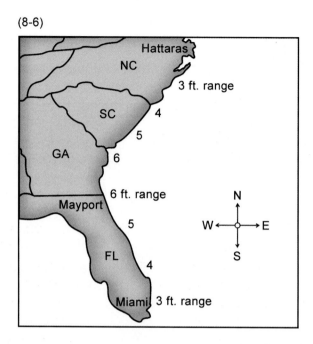

Near the center of this curvature, Mayport, Florida is almost 200 miles west of a line between Cape Hattaras and Miami. The tidal range at Mayport is about 6 feet, which is twice as great as the 3 foot tidal range that occurs at both Cape Hattaras and Miami. To a close approximation, the tidal range progresses mile by mile, from 3 feet to 6 feet, along this coast (see illustration above). Thus the geography of the coastline is the major influence on the range of the tide along this entire 800 miles of shoreline.

Now, we shift our attention from the contour of the coast, and consider the influence of the near shore topography, the slope of the seafloor over the continental shelf. The small oceanic tide remains relatively unchanged as it approaches a coast where the continental shelf is steep or vertical near the shore. See illustration 8-7(a).

(8-7) (a)

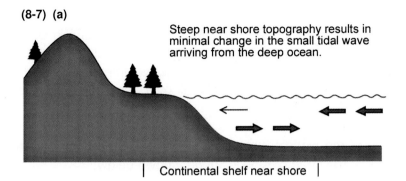

Steep near shore topography results in minimal change in the small tidal wave arriving from the deep ocean.

| Continental shelf near shore |

This is partly because some of the tidal energy is reflected back away from the coast by the steep continental shelf. Then, since the remaining tidal energy travels only a short distance to the coast, there are minimal shallow water hydraulic effects to amplify the tide. Conversely, **the tidal bulge is elevated if it approaches the coast over a shallow depth with a gradual slope that extends for many miles out to the continental shelf.**

(8-7) (b)

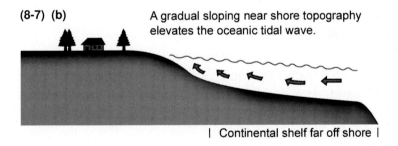

A gradual sloping near shore topography elevates the oceanic tidal wave.

| Continental shelf far off shore |

See illustration 8-7(b). This is because: (1) none of the energy is reflected at the shelf, and (2) since the tidal wave moves over a long shallow bottom, it is greatly amplified by the shallow water hydraulic effects that we detailed earlier in this chapter. As the tidal wave energy gradually moves into a shallower space, it is concentrated into a smaller volume of water, just as it is concentrated within a funnel shaped coastline. The only difference is that the concentration of energy into a smaller volume is horizontal into the receding coastline, and vertical over the gradually sloping bottom.

One of the world's highest tidal ranges is at Derby, Australia, which is at the head of King's Sound, shown below.

(8-8)

Ebb tide in King's Sound   Flood tide in King's Sound

One of the reasons for these 30 foot tides is the topography of the bottom. This basin has roughly parallel margins: it does *not* funnel inland. However, this wide, shallow basin has very gradual shoaling over a long distance from the mouth to the head of this broad basin (like a boat ramp). This concentrates the energy of the tidal wave into a smaller and smaller volume of water (shallower and shallower), resulting in an amplified tidal range.

The amplification of the deep ocean tidal range at the coast of the continents can be divided into two distinct stages. First, assuming that the continental shelf is not too steeply vertical, and the tidal energy is not reflected, the crossing of the shelf will amplify the 12–18 inch oceanic tide

by about twofold. This begins with slowing of the tidal wave as it crosses from about 1000 foot depth to about 300 feet, and the conservation of energy — less kinetic energy (slower wave) and increased potential energy (higher wave). This phenomenon is common to most of the continental coastline.

Second, the resulting 24 to 36 inch tidal wave moves into the coastal shallows and is further amplified by the contour of the shoreline and the slope of the near shore bottom, which is unique for each point along the coast. This can result in tidal elevation much greater than the three foot tidal bulge arriving from the continental shelf.

As part of the research for this chapter, I compared the tide tables for faraway shores with an atlas of geography, looking for a correlation between the contour of the coast with the range of the tide. Usually a receding coastline explained a large tidal range. Occasionally however, the correlation was inverse, so that the range was higher where I expected it to be lower. This was almost always due to the influence of the near shore bottom topography, which proved to be the dominant influence at that location. You must always consider both of these important parameters.

If the coastline recedes, this will elevate the tide. If it recedes and the bottom slopes slowly for a long distance offshore, the tide will higher still. If the coast recedes slightly but the continental shelf is vertical near the shore, the two parameters may negate each other and have little effect. **Here again, we have that pervasive central theme of tide patterns, tide tables, and this book: multiple independent simultaneous influences, such as the contour of the coast and the slope of the near shore bottom, may reinforce each other or may interfere with each other to create the unique tide at each location on the coast.**

In an interesting scientific footnote, the concept of tidal modulation by the coastal geography was confirmed by the arrival of a disastrous tsunami in Japan. Although accurate measurements of the event were impossible, it was noted that the same oceanic tsunami wave behaved differently after it moved into different bays and harbors. Furthermore, it was clear that the tsunami wave patterns were very similar to the patterns of the daily tides, which were also unique in each coastal basin. The tsunami was higher along the same coasts where the daily tidal range was greatest. Other tide patterns were also reproduced when the tsunami approached the coast over different near shore topography and into different coastal contours. This confirmed the theory that **coastal geography and topography were largely responsible for the variety of tidal patterns at different points of the same coast, all exposed to the same oceanic tidal wave**.

When yachtsmen and sailors take an extended cruise along any coast in the world, whether it is between the harbors of New England, down the Ivory Coast of Africa, or through the Scandinavian fiords, they will need to refer to their tide tables every time they pull anchor and get under way. It is likely that each segment of the coast will have somewhat unique tide patterns.

There are many interesting, complex, and unusual forces of nature that cause unique tide patterns. The vast majority of the time, however, the differences in the height of the tides along the way will be due to the constant variety in the contour of the coastline and the changing topography of the near shore bottom. There is an old saying in the medical profession: "When you hear hoof beats, don't think of zebras." In other words, a fever is more likely to be due to a cold, and not malaria. Likewise, a difference in the height of the tides is most commonly due to the most common difference along the coast, the contour of the coastline and the slope of the bottom.

Finally, these geographic features of the coastline also affect the time of arrival of the tides. In chapter six, we discussed how the tides may arrive at different points along a coast in a progressive pattern — later and later high tides as you proceed north or south, for example. Variation in the contour of the coastline and the near shore topography often contribute to this common finding. Since the tidal wave *slows down* in shallow water and its motion is retarded by a progressively narrow waterway, this affects the time of arrival of the tide along the coast.

Coastal geography and near shore topography may have another effect on the tides between the continental shelf and the waterline. This is a truly amazing and very important phenomenon known as "amplification by resonance." Resonance can be a powerful force. It is a fact that some sopranos can shatter a nearby crystal goblet by sustaining a note that resonates with the shape of the glass.

In a more scientific example, if you strike a tuning fork with the pitch "C" and hold it near a flute, the flute will begin to resonate and give off an audible humming sound *if* you are fingering the "C" note hole on the flute. It will *not* resonate if you are fingering the "E" note, or any pitch other than "C."

Finally, we have an example that can be explained by simple mathematics. If you pluck a guitar string, or draw the bow over a violin string, it is set in motion. See illustration 8-9. This motion has a waveform (~~~~~~~). The string has a certain length, and each wave (~) also has a measurable length. If you bring other motionless or "resting" strings into the room near the vibrating string, they may begin to vibrate under the influence of the sound. Whether they do, indeed, begin to move in a waveform depends on their length.

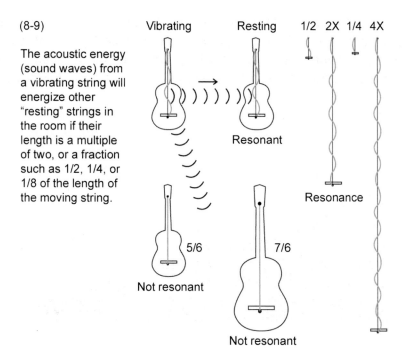

(8-9)

The acoustic energy (sound waves) from a vibrating string will energize other "resting" strings in the room if their length is a multiple of two, or a fraction such as 1/2, 1/4, or 1/8 of the length of the moving string.

A simple experiment will show that such resting strings will be set in vigorous motion (resonate) if they are: (1) the same length as the moving string, or (2) an even multiple of that length, such as twice as long, four times as long, or eight times as long, or (3) an even fraction of that length, such as 1/2, 1/4th, or 1/8th the length of the moving string. If the resting string is seven times the length of the moving string, or if it is 5/6ths as long, for example, then it will not resonate, because these are not even multiples (factors of two) of the length of the moving string.

Although it is a very rough analogy, we can begin our discussion of resonance in the ocean with a comparison to these musical instruments. In the coming pages we will look at the actual physical mechanism of resonance within a coastal basin.

But first lets use a little imagination, and say that as the oceanic tidal wave moves into shallow water, the "vibrating" ocean water comes in contact with the "resting" water of the coastal basin. Like the sound waves in illustration 8-9, the oceanic tidal waves moving toward the coastline have certain wavelengths from crest to crest. See illustration 8-10(a).

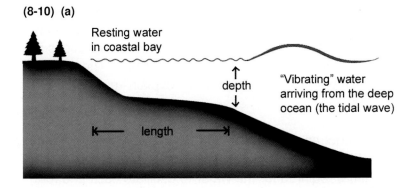

According to the laws of shallow water hydraulics, there is a mathematical ratio between the wavelength of the incoming oceanic tidal wave and the *length and depth* of the coastal basin, which will produce resonance in the resting shallow water. See illustration 8-10(b).

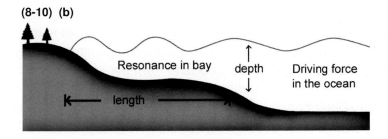

In other words, **if the length and depth of a coastal basin are attuned to the wavelength of the tidal wave arriving from**

the ocean, a new secondary resonant wave will form in the shelfwaters. If the coastal basin does not have dimensions attuned to the oceanic tidal wave, there will be no resonance. These secondary resonant waves are called "overtides," and they can generate the world's most extreme tidal ranges.

Before we proceed, I wish to remind the reader that when dealing with such complex scientific phenomena, verbal descriptions (without mathematics) are necessarily an approximation. The description in the above paragraph is a loose analogy, a helpful introduction, but incomplete. Therefore we will use other metaphors and models to explain resonance over the continental shelf.

The model above depends on the measurements (*length* and *depth*) of the coastal basin. The model below emphasizes the *period* (time) required for the water to oscillate back and forth from the mouth of the basin to the head of the basin and back to the mouth. This second model more accurately describes the motion of the tidal wave within coastal basins such as the Bay of Fundy, where extreme tidal ranges occur.

The coastal basin may be regarded as an elongated rectangular pan of water which is open on the end toward the ocean, the "mouth" of the bay. If you briefly lift one end of a pan of water and then return it to a level position, a wave will oscillate back and forth over its length. In a coastal bay, the water is set in motion at the open end (mouth) by the energy of the incoming oceanic tidal wave. It will then rock back and forth from the mouth of the bay to the "head" of the bay (farthest inland extension) and back to the mouth. Each coastal basin has a natural period of oscillation: a proscribed time required to propagate one cycle of oscillation, twice over its length.

The critical measurements that determine the natural period of oscillation in a bay are: (1) the depth of the basin at the mouth, and (2) the length of the basin from the mouth to the head of the bay. **If the natural period of oscillation within a coastal basin is synchronized with the period of the tidal wave arriving from the ocean, then the water in the bay will respond with more vigorous oscillation, and there will be a much greater tidal range at the head of the bay.**

This oscillation of tidal energy within a coastal basin is the same phenomenon as the seiche effect within oceans and gulfs described in the preceding chapter. We have seen how the seiche effect can cause interference of tidal wave constituents in some basins such as the Gulf of Mexico, and how this interference of reflected waves and incoming waves can cause some constituents to lose their influence within such basins. We have also seen that by the interaction of oscillating waves over the continental shelf, these waves can reinforce each other, and generate a standing wave, which raises the water level near the coast. Now in this chapter we will learn that seiche effect oscillating waves within a coastal basin, interacting with the incoming energy of the oceanic tide, can create the world's most extreme high tides.

In order to become familiar with the basic concepts and terminology involved, illustration 8-11(a) depicts a wave oscillating within a closed basin such as an inland lake. The force that generates this tide-like motion is usually a rapid change in barometric pressure over one end of a large elongated lake. This is the phenomenon first studied by Laplace in Lake Geneva, Switzerland, and later applied to understanding the dynamics of coastal tides. When this sudden change in pressure occurs at one end of the lake, it makes the water level uneven, similar to the effect when you climb out of the bathtub. The water then oscillates back and forth in the closed basin. It is alternately higher at one end and then the other, and there is a midpoint in this oscillation called the "node." In a *closed* basin, the wavelength of this oscillation will be twice the length of the lake.

To put it another way, the length of the lake will be one-half of the wavelength of the oscillation. Such a closed basin is called a "half-wave oscillator."

(8-11) (a)

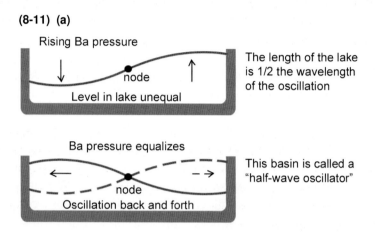

In the case of an *open basin* such as a coastal bay, the force that generates the motion is the rising and falling of the oceanic tide at the mouth of the bay. In illustration 8-11(b), the ocean tide has a wavelength and a certain period of time between the crests. If the length of the coastal bay is exactly 1/4th of this wave-length, the node of the shallow water oscillation will be positioned exactly at the mouth.

(8-11) (b)

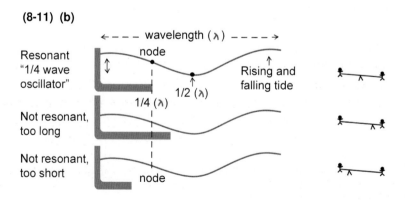

Such a basin is called a quarter-wave oscillator. Compare 8-11(a) with 8-11(b), and notice the similarity between the node in the center of the tide-like motion in the lake, and the node at the mouth of the coastal basin. When the length of the coastal basin positions the node of natural oscillation at the mouth, the period of the basin is synchronous with the period of the ocean tide, and there will be a "sustained forcing" of the tidal oscillation within the bay. This is analogous to a child's see-saw, which allows an efficient transfer of energy when the pivot is exactly in the center.

If the coastal basin is too long, or too short, the node of oscillation is not at the mouth of the basin. In this case, the natural oscillation within the bay is not synchronized with the ocean tide, and the forcing of the coastal water will not be sustained, just as the see-saw doesn't transfer energy efficiently if the pivot is not in the center.

The metaphor of a child's swing can provide additional insight into the mechanism of sustained forcing. Remember that the period is a measure of *time*. Little children can increase the height of their motion on a swing only if they push off the ground or pump their feet in perfect timing with the period of their motion. If you are pushing a child on a swing, you must time each push perfectly in order to add to their height and motion. If you push too early or too late, you will provide plenty of energy, but you will impede the action that you want.[c]

In just the same manner as you add more and more energy to a swing by pushing repetitively in just the right rhythm (the same rhythm as its established motion), the tidal oscillation within a coastal basin *will gain more and more energy from successive tidal waves* arriving at the mouth, if the period of oscillation in the basin is synchronized with the period of the forcing tidal wave.

[c] You can also add energy to a child on a swing if you push on every other cycle, or every fourth (or even, every eighth) cycle of their swinging motion. This is reminiscent of the resonant guitar strings that started this discussion.

We learned in the previous chapter that all basins experience a seiche or oscillation once the water is set in motion. All coastal bays and inlets have these seiches, with reflected waves oscillating back from the head to the mouth. However, only those bays and inlets that have the right length (1/4 of the ocean tidal wavelength) and depth (to create a period equal to the tidal wave period) will have a period of oscillation that allows *sustained forcing* of more and more tidal energy into the coastal basin.

Thus, **the tidal waves arriving from the ocean will provide sustained forcing of the oscillation within a bay only if the natural period of the coastal bay is synchronized with the period of the forcing tidal wave, so that each successive oceanic tidal wave can add more and more energy to the coastal water in the same rhythm as its natural motion. By this remarkable balance of forces, the world's highest tides are generated.**[d]

Sustained forcing adds more and more energy into the coastal bay with each successive ocean tide. This raises the question of why sustained forcing doesn't elevate the tidal wave ad infinitum; what prevents an endless accumulation of tidal energy in the bay? There are four answers. (1) There is friction between the water and the shoreline, and friction within coastal estuaries (see chapter 9). (2) The earth's rotation diverts the path of the waves in and out of the coastal basins, which ultimately results in an inefficient transfer of energy. (3) Some of the energy is absorbed by the somewhat elastic planet, and (4) there is loss of energy from the system when tidal energy is reflected back from the head of the bay to the open mouth and then back into the ocean. After eons of time and countless successive tides

---

[d] See supplement at the end of chapter eight.

force energy into these bays (and some of the energy is lost), the system reaches an equilibrium with a stable pattern of tides.

The most extreme tidal ranges due to this phenomenon are always located at the head of the bay, the point farthest from the ocean. In the Bay of Fundy, the Minas Basin has 45 to 53 foot tides at the point farthest from the mouth of the Gulf of Maine. Notice in the picture on page 123 that the tides are 18 feet at Grand Manan near the mouth, 33 feet at St. Martins halfway up the bay, and 50 feet at the head of the Bay of Fundy. In another example, Turnagain Arms, Alaska, has 30 foot tides because it is the point farthest from the mouth of Cooks Inlet. Now picture the children on the see-saw one last time. The pivot of the see-saw is analogous to the node of the oscillation at the mouth of a coastal basin. The wave of oscillation rocks up and down within the basin so that the highest water level is at the head of the bay, the point farthest from the mouth, just as the children go highest on the see-saw because they are at the point farthest from the pivot.

Sustained forcing of the resonant co-oscillation in the Bay of Fundy is a major reason for the enormous tides in this place. Mathematical formulas define the critical length and depth of a coastal basin, as well as its natural period. It is not necessary to know these formulas[e] in order to appreciate the following:

(1) A depth of 550 feet is ideal for resonant co-oscillation between a basin of this size and the semidiurnal tide over the continental

---

[e] The formula for the generation of resonance and sustained forcing in a coastal basin is: Period of oscillation in the coastal basin = 4 times (the length of the basin) divided by the square root of (the depth times gravitational acceleration), or **T (time) = 4L /the square root of dg**. This was derived from a very important formula (Merian's formula, 1828) that describes the oscillation in a closed basin such as a lake: Period of oscillation = 2 times length/(the height of a wave above the standing water level) divided by (the square root of depth times gravitational acceleration), or **T (time) = 2L / (h / the square root of dg)**. In turn, this formula was derived from a basic formula in oceanography (Lagrange, 1781) for the velocity of a wave in deep water: velocity = the square root of (the depth times gravitational acceleration), or **v = the square root of dg**.

shelf at the Gulf of Maine. The depth at the mouth of the Bay of Fundy varies between 400 and 700 feet.

(2) The ideal length of a bay for sustained forcing by the Gulf of Maine semidiurnal tide is 177 miles length, which is 1/4th of the semidiurnal tidal wavelength over the continental shelf at the mouth of the Bay of Fundy. The actual length of the Bay of Fundy is 180 miles.

(3) The period of time for the tidal wave within the Bay of Fundy to travel from the mouth to the head and back to the mouth is about 12 hours, 40 minutes. This is very close to being perfectly synchronized with the period of the Atlantic semidiurnal tide, which is 12 hours, 26 minutes.

All of this means that: **(1) the enormous tides in the Bay of Fundy would not occur if the depth was deeper or shallower, or if the length of the bay was significantly longer or shorter than its actual length — within 2% of the ideal length, and (2) the extreme Bay of Fundy tides would not occur if the period of natural oscillation of the bay was significantly longer or shorter than its actual period — almost exactly equal to the Atlantic semidiurnal period. These are perfect conditions for amplification by resonance and the sustained forcing of the natural oscillation in this bay.**

Shallow water resonance is so important that NOAA includes this influence in the computer program for predicting the tides. It does this by adding the influence of the resonant overtides to the other constituents that contribute to the tide along the coast. The principal lunar semidiurnal constituent of the tide is called M2. The resonant overtide with 1/2 the semidiurnal wavelength that occurs over the continental shelf is called M4. The overtide with 1/4th the semidiurnal wavelength, that may occur when M4 enters shallower coastal basins, is called M6. The overtide, M8, has 1/8th the wavelength of M2, and may occur when M6 enters still shallower waterways.

The most extreme tidal range on earth is at the head of the Bay of Fundy. The torrent of water into this basin raises the water level fifty feet in only six hours. Each tide brings a volume of water greater than all the rivers on earth combined. Every twelve hours, each flood adds over 100 billion cubic meters of seawater to the bay. The weight of the flood tide causes a measurable lowering of the adjacent landmass, and the edge of the continent rises back up with each ebb tide.

(8-12)

Low tide in the Bay of Fundy.

Waterline is at bottom left.

This oceangoing ship is high and dry against the dock.

In only 6 hours, this ship will float out on the high tide.

A major cause of the Fundy tide is resonant co-oscillation and sustained forcing, also known as amplification by resonance. Another important factor is compression of a large amount of tidal energy from the Gulf of Maine into the funnel shaped receding coastline of the Bay of Fundy. There are other coastal basins with the right dimensions to promote sustained forcing and 30 foot tides. Other basins have an extremely funnel shape and 30 foot tides. Fundy's tides are even greater because it has both: not only ideal dimensions for amplification by resonance, but also a dramatic funnel shaped receding shoreline.

Among the greatest tidal ranges are: The Bay of Fundy, Nova Scotia — Cooks Inlet, Alaska — Bristol Channel, England — Kimberley Coast, Australia — Okhotsk Sea, Russia. These places all have tidal ranges of 30 to 50 feet. They are scattered randomly around the globe. They occur in both the Atlantic and Pacific oceans, and are just as likely to be near the equator or near the poles. We have just discussed the geographic and hydraulic influences that cause the Bay of Fundy tides. Resonance and sustained forcing of the natural period of oscillation is also the cause of the 30 foot tides at the head of Cooks Inlet.

The extreme tides in the English channel are so complicated that entire books are dedicated to this subject. They are due to a combination of a coastal Kelvin wave (see page 84), M4 and M6 overtides (page 138), and resonant co-oscillation — the channel is approximately a half wave oscillator with the North Sea tidal wave (page 134). You could think of the English Channel as a long narrow lake with the node of its natural oscillation in the middle, except that both ends are open to the tidal waves from the North Sea and Atlantic Ocean.

The 30 foot tidal range at Derby Australia is chiefly due to the contour of the bottom topography in King's Sound, plus the fact that this entire coast of Australia receives tidal wave input from both the Pacific Ocean and the Indian Ocean, converging at the Kimberley Coast. Remember that the Indian Ocean is west of Australia, and the only way that these tides could influence the Kimberley coast of Australia is because of the horizontal tractal force of gravitation gathering the Indian Ocean toward the east, while it gathers the Pacific Ocean toward the west.

Of all of the extreme tides scattered around the globe, the Okhotsk Sea, Russia, has some of the most complicated tidal dynamics. The entire basin is located on a very wide and shallow shelf of this continental coast, which facilitates a large tidal range. This sea is a semi-enclosed basin, bounded on the west by the coast of Russia. It is partially bounded on the east by the Kuril Islands. These are a series of small islands between the Sea of Okhotsk and the Pacific Ocean, shown below. A seiche establishes a secondary oscillating tidal wave within the sea. This oscillation would ordinarily dissipate quickly because this is a relatively small and shallow basin.

(8-13)

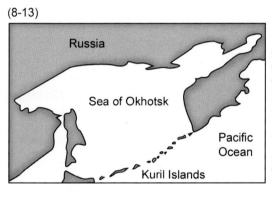

The tidal energy from the Pacific Ocean enters the Sea of Okhotsk through the 13 straights between the chain of Kuril Islands.

The entire sea is over a very shallow continental shelf.

There are thirteen separate straits between the string of Kuril Islands. Some of the straits are the approximate depth that may facilitate sustained forcing of the natural period of oscillation within the sea. Some of the straits do not facilitate sustained forcing, because they are not the appropriate depth. There are other complex hydraulic considerations that contribute to the tide pattern within the Sea of Okhotsk. Within this basin you may find a tidal range of 30 feet or a tidal range of only three feet on different shores less than 100 miles away. This is a wonderful example of how geography and shallow water hydraulics shape the global tidal wave into a variety of patterns.

As we have just seen, at the margin (mouth) of a coastal body of water, such as the Gulf of Maine or the Mediterranean Sea, the energy of the adjacent ocean "co-oscillates" with the water within the coastal basin. The amount of energy exchanged between the ocean and the gulf, bay, or sea depends on: (1) the ocean wave characteristics, (2) the length and depth of the basin, and (3) the natural period of oscillation of the coastal basin. There is another important parameter that determines the strength or amplitude of the co-oscillation in any coastal basin — the width of the mouth. See illustration 8-14.

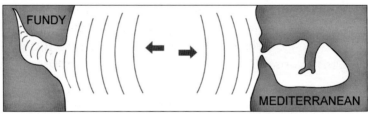

You don't have to be an oceanographer to see that when the mouth of a coastal basin is wider, the force of the oceanic tidal wave admitted into the basin will be greater. This is a major reason why there are minimal tides in the Mediterranean Sea (very small mouth at the Straits of Gibraltar), and huge tides at the Bay of Fundy, in the Gulf of Maine (very wide mouth to the Atlantic). **The width of the mouth of a coastal basin determines the amount of co-oscillation between the ocean and the bay, and determines the amount of energy forced into the bay. This is a major reason why diverse gulfs and seas have such different tidal amplitudes and patterns.**

An additional reason for the complexity of coastal tide patterns is the possibility of "independent tides" arising within the largest coastal basins. Landlocked lakes have independent tides, forced by lunar and solar gravitation on the water confined to these inland basins. They are rarely significant or even noticeable, although the great lakes at the southern border of Canada have independent tides of about 1½ inches. In order to have independent tidal oscillation, a lake or a coastal basin must be long enough so that there is a significant difference between the gravitational effect on one end compared to the other. The basin will not be set into oscillation just because the water weighs more (or less) throughout the entire lake or coastal bay.

Large coastal basins may have independent tides (determined by the direct gravitational forces on the water within that basin), in addition to the common co-oscillating tides (caused by the driving force of the oceanic tidal wave at the mouth of the basin). The admixture of these two tide patterns may contribute to even greater complexity of the tides in some coastal gulfs, bays, and harbors.

Supplement to chapter eight:

In chapter seven, we discussed the hydrodynamics of the Gulf of Mexico (pages 103–104) that are very similar to the mechanism of sustained forcing (pages 135–136). Recall that the seiche in that gulf has a period that is close to the 24 hour period of the global diurnal tidal wave. We said that this interaction between the internal wave (the seiche) and the external forcing wave supports a diurnal pattern within the Gulf of Mexico.

However, the Gulf of Mexico has a *small* tidal range, not a large tidal range that might be expected from sustained forcing. So, why doesn't sustained forcing generate extreme tides in this gulf? We now have enough information to explain this apparent inconsistency. First of all, the period of the seiche in the Gulf of Mexico is not close enough to the global diurnal wave period to assure sustained forcing. We might call this "near-resonance." Secondly, the mouth of the Gulf of Mexico is relatively narrow, and not enough forcing energy can enter this basin with each successive diurnal wave from the Atlantic ocean. Finally, the diurnal tidal influence is only half as strong as the semidiurnal tidal influence (chapter four). For all of these reasons, sustained forcing of the diurnal pattern of tides does not occur, in spite of the fact that the internal seiche is roughly synchronized with the global diurnal tidal wave.

There is another interesting consideration. The oceanography literature refers to the possibility that the natural oscillation of the *entire planet earth*, which is somewhat elastic, appears to be closer to 12 hours than 24 hours. Some geophysicists point to this as an explanation of the fact that semidiurnal tides are most common around the globe. If this seems a little far fetched, consider the following. All mechanical systems have a natural period of oscillation. That is why a clock can be regulated by a swinging pendulum of a certain length and weight. This is also true of an elastic band or a coil spring stretched by a lead ball, a trampoline bouncing a large man, or a planet orbited by a large moon.

When I first read this theory, I thought of a lecture that I once heard on the physiology of respiration (breathing). When you see a dog "panting" with very rapid breathing in order to cool down, you are looking at the natural period of oscillation of the (relaxed) muscles in his chest wall. In order to conserve energy while he pants, the dog sets up an oscillation of the completely relaxed muscles surrounding his lungs, which can then spring back and forth without exerting much energy. My dog pants at six shallow breathes per second. This is the natural period of oscillation of his chest.

It is not hard for me to accept that semidiurnal oscillations in the oceans and coastal basins are more "natural" on the earth, and this may be part of the reason that semidiurnal tidal patterns are common, while diurnal patterns are rare.

Chapter Nine

Shallow Estuaries and Tidal Bores

"There is a tide in the affairs of men, which, taken at the flood leads on to fortune. Omitted, all the voyage of their life is bound in shallows."
                                                                Shakespeare

    If you took a poll of every knowledgeable science historian, their list of the ten greatest physicists of all time would invariably include Neils Bohr. A legend in his own time, Neils Bohr was one of a small group of theoretical physicists in the same league with his contemporary, Albert Einstein. After Bohr won the Nobel Prize, a young American physicist made a pilgrimage to visit his laboratory. On entering the lab where the mysteries of quantum mechanics were solved, the young man noticed a horseshoe nailed to the wall over Bohr's desk. He couldn't imagine that Bohr's promethean brain could be influenced by a lucky charm, and he had to ask the great man if he believed in the horseshoe. "Of course not," Bohr said, "but I understand that it works whether you believe in it or not."

    My research on the tides was plagued by one baffling subject: the unique tide pattern seen in estuaries. The unique pattern of ebb and flow in estuaries is described clearly in the literature, but I couldn't find any scholars who offered an opinion on *why* these patterns emerged there. Most textbooks on the subject are satisfied to describe the estuary tide pattern, and then ask us to take it on faith that someone, somewhere, understands why this pattern occurs.

There are still a few subjects in science that are essentially incomprehensible to everyone who has studied them. The best example is turbulence of fluids and gases, a subject that almost defies scientific analysis, a subject that gives physicists nightmares. Einstein took up the problem of turbulence late in life, but it made him irritable and gave him headaches, so he left it unsolved. It is interesting that estuaries involve fluids — turbulent fluids.

Shallow estuaries have a singular effect on the tide cycle, because there are unique dynamics of flow within these backwaters of the tidal zone. In shallow estuaries and tidal flats there is friction between the water, the seafloor, and the shoreline. This friction is related to the temperature and viscosity of water, the slope and structure of the bottom, the contour of the shoreline, and the movement of the water through vegetation, oyster beds, and man-made structures such as docks and bridges. All of these parameters cause significant effects on the tide in estuaries, which do not occur on the coastline.[a]

All of these parameters are complex and irregular. Picture the infinitely irregular shoreline of such a basin, lined with oyster bars and vegetation. Recall the endless bumpy contour of the bottom seen on your depth sounder. These surfaces can only be described as "irregularly irregular." It is very doubtful that anyone will ever devise a computer model of an estuary that can recreate such a system. Many of the parameters are constantly changing: currents erode the shoreline, silting and sedimentation alter the bottom, the vegetation is seasonal, the water itself is changing temperature and viscosity, which are dependent on freshwater run-off.

---

[a] In the calculation of the NOAA tide tables, 37 constituents are used for most tide monitoring stations along the coast. However, some of the stations located up rivers and near estuaries require up to 114 constituents to account for the deformity of the tide cycle. explained in this chapter.

The viscosity of a liquid is a measure of its resistance to flow. Honey has greater viscosity than water. Cold honey has greater viscosity than warm honey. The viscosity of water is very significant. If you tilt a bucket of water at different angles and measure the time required to empty all of the water, the bucket will empty faster and faster as you increase the angle of tilt. However, there is a limit to how fast any such container can be emptied. There comes a point where the time required to empty any container does not change, no matter how much you increase the angle. The viscosity of the water alone determines the upper limit of flow from a basin, no matter whether it is a bucket of water or an estuary.

During a severe rainfall event that caused flooding of the upper Mississippi River, the Corp. of Engineers was asked to provide predictions of the time of the highest flood level at various towns along the river. This was important so that the townspeople could be ready for the peak water level, build dikes, or evacuate at the appropriate time. Unfortunately, some predictions were incorrect for locations down river from the rainfall. After the event was over, the engineers met to investigate the source of this error. Their computer models were almost entirely based on the pattern of peak water levels along this same river during previous floods. It turned out that many bridges had been built since the last major flood. The viscosity of the water flowing through the pilings of these new bridges caused a significant delay in the movement of the flood down river. Of course, the influence of viscosity and friction is much greater within a vegetated estuary.

Coastal engineer, Dr. Kevin Bodge, suggested to me that the best way to understand the flood and ebb in such a place is to think of a sponge. It is easier to fill a sponge with water by the force of a faucet than it is to drain the water from a sponge by gravity. The flood tide is driven into an estuary by the astronomical forces that lift the oceans and send them rushing toward the continents. The ebb from an estuary is water draining back toward the resting level. **These dynamics may cause a distortion of the tidal waveform within an estuary, so that the falling tide requires more time than the rising tide.**

Since the flood occurs more quickly than the ebb in these places, the floodtide *current* is stronger than the ebbtide current. This has a significant effect on the sedimentation and silting of such waterways. The faster floodtide currents transfer more sediment inland than the ebbtide current carries out. This is a major consideration for hydraulic engineers who must deal with problems associated with sedimentation. Over long epochs of time this process changes the physical parameters of such coastal basins, and ultimately changes the tide pattern.

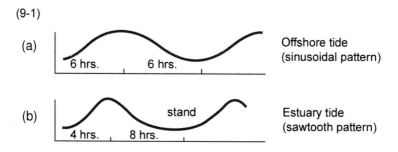

Illustration 9-1(a) depicts the symmetrical sine wave pattern of an equal flood and ebb which is usually found in open waters and along most coastlines. However, because of the viscosity of seawater and the friction within shallow vegetated estuaries, this sine wave may be distorted along the coast, yielding the sawtooth pattern seen in illustration 9-1(b). The twelve hour tide cycle in an estuary might consist of a 5½ hour flood and a 6½ hour ebb. In more extreme conditions (larger, shallower basin, with more obstructions) the tide cycle might consist of a 4 hour flood and an

8 hour ebb. Illustrations 9-1(b) and (c) depict the sawtooth pattern of the tide cycle, where an estuary has elongated the ebb. The flood tide pattern is deformed because it arrives while the previous tide is still flowing out of the basin — the rivers are still flowing out to sea when the next tide starts upriver. The flood is more abrupt, steeper and quicker. This deformity of a flood tide that flows against the previous ebb tide can be described by advanced mathematics; but we will simply have to accept the notion that a wave is bound to be effected if it is moving upstream. It seems common sense that an elongated ebb tide would condense or compress a flood tidal wave which is trying to enter a basin against the current.

(9-1)

(c)

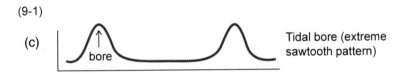

Tidal bore (extreme sawtooth pattern)

One interesting consequence of this tidal waveform deformity is known as "tidal pumping." **Because the outgoing ebb requires more time than the incoming flood, some of the water is still within the estuary, bay, or river when the next tide cycle begins. With each successive tide, there is more water in the basin.** This tidal trapping of water results in elevation of both the high tide and the low tide levels. See illustration 9-2(a). When you put a month of these successive tides on a graph, you see that the "half tide level" (a line connecting the halfway point between the high and low tides) rises higher and higher around the full moons and new moons, and falls back lower at the quarter moons. This is because the force of the flood is greater during spring tides than at neap tides, and there is more tidal pumping and trapping during spring tides.

This pattern of rising and falling half tide levels is called the "fortnightly tide pattern." Illustration 9-2(b) depicts a generic tide cycle near the coast where there is no estuary effect. The fluctuation of the tide cycle in the ocean and on the coast is dominated by the lunar influences. In basins with semidiurnal tides, **in deep water and on most coastlines, higher high tides are usually associated with lower low tides, because the dominant lunar component of the tide must displace the global oceans in this way.** When a larger volume of water is displaced toward the lunar tidal bulges, there is a smaller volume of water available between the bulges, causing lower low tides on days with higher high tides.[b]

(9-2)

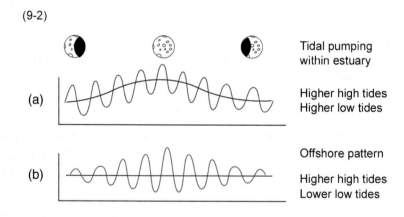

[b] It is true that the lunar tidal ellipse displaces the oceans so that there will be lower low tides associated with higher high tides. This will be the generic tide pattern unless the estuary effect and tidal pumping are strong local influences. However, the model that is most helpful to explain all of the variations in high and low tide levels seen throughout the year, is as follows: (1) for a given date, consider the lunar parameters (distance, declination, phase) and establish whether the "lunar tides" will have high "highs" and low "lows," or moderate "highs" and "lows." (2) Consider the strength of the solar tidal influences (distance, declination, thermal expansion) on that same date. (3) During spring tides, *add* the solar "high" to the lunar "high," and add the solar "low" to the lunar "low." (4) During neap tides, *subtract* the solar "high" and "low" from the lunar 'high" and "low," respectively. This method will explain almost all variations on all dates (see pp. 236–237).

Table 9-2(a) is for the inshore waters. **In estuaries, tidal pumping may cause cycles of fluctuating half-tide levels, resulting in a linkage between higher high tides and higher low tides.** This deformity of the tide cycle, tidal pumping, and the fortnightly pattern are so significant that NOAA includes them in the computer program used to generate the tide tables.

The deformity of the tidal waveform seen in estuaries may extend into the rivers that fill and drain these inland waters. To a lesser degree, it may also extend into other coastal waters, adjacent to estuaries. This results in some unusual and interesting tide patterns. For example, with an extreme elongation of the ebb tide, as seen in 9-1(b), the ebb tide portion of the tide cycle may be completely flat. This is known as a "stand" in the tide, with no change in water level for many hours.[c]

Another feature of the tide tables due to the estuary effect (tidal trapping) is shown in the sample tables of "Tide Corrections" shown below. It is common practice in publications that include the entire calendar year of tide cycles, to print the tide table for a major port, and then offer the "Tide Corrections" for all other nearby ports. In the example below, corrections in the left column refer to the tide tables for Mayport, Florida, which is at the mouth of the St. Johns River, in northeast Florida. The Tide Corrections list includes Amelia City (high tide correction of 0.42 hours), Nassauville (correction of 0.52 hours), and Palatka, Florida (correction of 7.14 hours). Of course, these corrections mean that if the high tide is predicted for 12:00 noon at Mayport, the high tide

---

[c]There are other causes of a "stand" in the tide with no change in the height of the water for a period of hours, occurring between the high and low tide. On page 69, there is a discussion of the interference of two tidal constituents, which may completely cancel each other out, resulting in a "flat" tide table. The summation of the many tidal wave constituents can and does occasionally result in a stand in the tide along the coast. On page 105, we learned that over the continental shelf a reflected wave toward the east can mix with the following tidal wave toward the west, and create such a stand. Finally, in places with minimal tides, on days when there is only gradual tidal forcing from the ocean, the tidal movement may be impeded by an opposing wind.

will be at 7:08 p.m. at Palatka, for instance. The major reason for the delay in arrival of high tide at these ports is their distance inland from the coast. Amelia City is a few miles inland, Nassauville is about one mile further from the coast, and Palatka is about 50 miles upriver from Mayport. This delay simply reflects the time required for the tidal wave to move up into the progressively narrow waterway.

(9-3)

Tide Corrections

Mayport	HI	LO	Key West	HI	LO
Fernandina	+0.20	+0.14 hrs	Duck Key	−1.11	−0.40
Chester	+0.37	+0.39	Sombrero Key	−1.01	−0.38
Amelia City	+0.42	+1.01	Flamingo	+5.35	+7.38
Nassauville	+0.52	+1.35	Dry Tortugos	+0.35	+0.40
Jacksonville	+1.12	+1.48	Cape Sable	+3.36	+4.48
Orange Park	+3.37	+4.48	Chatham	+3.32	+4.46
Green Cove	+5.14	+6.11	Shark River	+3.20	+4.48
Palatka	+7.14	+8.19	Lostmans River	+3.22	+4.22

The interesting thing is that the delay in arrival of low tide is longer than the delay of high tide at each inland location. At Amelia City, the correction for high tide is 0.42 hours, and the low tide correction is 1.01 hours (0.59 hours greater delay for low tide). At Nassauville the correction for high tide is 0.52 hours, and the low tide correction is 1.35 hours (0.83 hours greater delay). At Palatka the high tide correction is 7.14 hours, and the low tide correction is 8.19 hours (1.05 hours greater delay). Amelia City and Nassauville are both located in the great saltmarsh estuary of the southeast coast, comprised of millions of acres of flooded spartina grass. Palatka is on the St. Johns River. Home to many alligators, the St. Johns is a slow moving, black water river, with swampy banks and innumerable small tributaries which flood cypress trees inhabited by soft shell turtles and water moccasin snakes. This vegetated waterway and these grass filled salt marshes are exactly what Dr. Bodge had in mind when he devised his analogy of a sponge which distorts the tide cycle.

Corrections are also shown for the Key West tide station. Notice that the correction for the *mouth* of Shark River is 3.20 hours for high tide, and 4.48 hours for low tide. The corrections for the mouth of Lostman River are 3.22 hours for high tide, and 4.22 hours for low tide. These places are at the coast of southwest Florida, not for a port that is 50 miles inland. There is an extreme deformity of the tide cycle that causes a greater delay of low tide on the coast, at the mouth of these rivers, because they flood and ebb within the Florida everglades. You will not find this severe deformity of the tide cycle along the rocky coast of Maine. **The amount of deformity of the tide cycle due to tidal trapping, and the extent to which it influences the adjacent coast, depends on the nature of the waterways in the inshore tidal zone. There is a greater influence where shallow vegetated estuaries are present. Because of this tidal trapping, there is often a greater delay in the ebb tide correction than the flood tide correction as you proceed inland.**

In the example above, the arrival of high tide is progressively later as you proceed upstream, away from the coast. This is the case in virtually every waterway. Furthermore, it does not matter if the waterway turns back 180 degrees toward the coast — the tide arrives later at each port farther upriver from the ocean, not at each port farther west (or east on the west coast).

As regards the height of the tide proceeding up coastal rivers and bays, the pattern is less certain. Most coastal waterways have lower and lower tides as you progress away from the ocean, because of the loss of energy from the system. However, it is not really surprising to find the opposite pattern. Recall that one of the reasons for the extreme tides at the Bay of Fundy is sustained forcing of the natural oscillation of high waters from the head of the bay to the mouth and then back to the head. In all places where this occurs, the tidal range is always highest at the point farthest from the mouth — we likened this to a child at the end of a see-saw, the point farthest from the pivot.

The two largest rivers on the coast of Maine are the Kennebec, which has the more common pattern of *lower* tides along most of its course upriver, and the Penobscot, less than one hundred miles away, in which the coastal section has generally *higher* tides progressing inland. **The unique length and depth of each coastal waterway determine whether the tidal range decreases or increases as you get farther from the ocean.**

When there is an extreme example of the pattern, as in 9-1(c), the very steep slope of the flood tide portion of the tide cycle may create a violent "tidal bore," with a foaming, churning wall of water racing up a river. Prolongation of the ebb tide contributes to this phenomenon, since bores occur where the river is still flowing out to sea when the rapid flood arrives with the following tide cycle. As previously noted, this produces an abrupt flood tidal wave. In addition, when the river or bay has both progressive narrowing of the shoreline and a long gradual shoaling of the bottom, the leading edge of the flood tide gains height as it loses speed. Finally, it develops a steep, collapsing wavefront, the hallmark of a tidal bore. Behind this leading wave, there may be a series of high frequency, smooth surface waves.

This is not a common result of the estuary tidal wave deformity. Tidal bores are rare and dramatic floods that occur under unique circumstances, at about 60 places on earth. They may occur: (1) within 4 days of the spring tide in places that (2) have a tidal range of at least 20 feet, (3) have a deformity of the tide cycle, with very abrupt rise of the flood tide phase, and an elongated ebb tide phase, (4) have persistent current out to sea at the time the following tide begins to flood, (5) have progressive shoaling of the bottom as the basin progresses inland, and (6) have a progressive narrowing of the basin toward the head.

An amazing tidal bore occurs in the Amazon River basin, where on occasional days the tide races up the river in a wall of water 10 feet high, traveling 15 mph. This bore travels up the river 300 miles. The roar of this deluge can be heard 15 miles away. The rivers of the Minas Basin at the head of the Bay of Fundy produce tidal bores that are not as high as some others, but these three foot high tidal waves occur much more often, coming every month, near the spring tides. The local tourism board advertises that you can, "hear the pull of the moon," at these times.

(9-4)

15 ft. high wavefront of the tidal bore in the Qiantang River, China. This bore travels 300 miles inland.

The deluge of the tidal bore racing up the Amazon River can be heard 15 miles away.

In the Fu Ch'un River estuary in China, the world's most dangerous tidal bore may attain 20 feet in height, and travel upriver at 15 miles per hour. After an arduous voyage from England to the South China Sea, a captain Moore of H.M.S. Rambler secured a safe harbor at the mouth of another river in

China, the Tsientang Chiang River. He had been forewarned that there would be a tidal bore during his anchorage there. On the appointed date, he had his ship double anchored and instructed his officers to have the engines running against the tide at full throttle. In spite of this preparation, his transoceanic ship was carried, out of control, up the river for three miles.

(9-5)

These surfers on the La Dordogne River, France, are riding the moving surface swells that follow the tidal bore wavefront.

I am sure that you have heard of groups of amateur astronomers traveling to another continent to witness an eclipse of the sun. There are also "tidal bore surfing clubs" that jet around the world to ride a surfboard up a river on the next tidal bore. Shown above are surfers on the La Dordogne River, in France. They are riding the moving surface swells that follow the tidal bore wavefront.[d]

---

[d] My wife has her own definition: a tidal bore is a retired husband who sits around the house writing a book about the tides (and to think, I *was* going to dedicate this book to her).

Chapter Ten

Computation of the Tide-Tables and Chaos Theory

"To measure is to know."   Johannes Kepler

The Fleisher–Harris tide predicting machine is shown below in illustration 10-1. This amazing piece of engineering utilizes many gears and camshafts, each of which represents one of the celestial orbits or other tide table constituents. It is able to account for 37 independent cyclic forces mechanically.

(10-1)

This 2500 pound, 11 by 6 foot machine utilizes many gears and camshafts to represent 37 celestial cycles and other tidal influences.

This amazing piece of engineering was used at NOAA to produce the tide tables, before the age of computers.

This ingenious device was used by NOAA to generate the tide tables for decades before the age of computers.

The history of tide predicting machines in the past century is a microcosm of recent technological progress. In 1833, the British Admiralty produced the first tide table for the fleet of her majesty's ships. The calculation and production of the tables were entirely by hand, although a form of Gutenburg press was available for distribution. In 1873, The British Association for the Advancement of Science commissioned the construction of a mechanical "integrating machine to compute the height of the tide," which is shown below.

(10-2)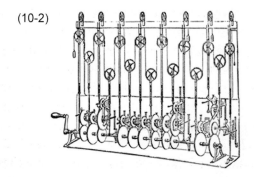

Tide predicting machine used in 1876.

The first tide predicting machine in the United States was designed by William Ferrell of the U. S. Coast and Geodetec Survey. In 1882, this device began to successfully predict the time of high and low tides, but it did not provide any intermediate tide heights. This was replaced by the Fleisher–Harris machine at NOAA, in 1910. See illustration 10-1. Based on advanced mathematical formulas and 37 tide constituents, it produced a graphic tracing of the height of the water throughout the tide cycle for any point on the coast, if it was provided with very labor intensive manual attention. Using the data from the Fleisher–Harris machine, the NOAA tide table master pages were hand typed each year, up until 1965.

In 1966, large mainframe electronic computers, running tide prediction software written in FORTRAN, took over the calculations. Throughout the 1970's, large IBM computers generated a closet sized room full of 80-column computer punch cards, which required extensive hand sorting, until a computer-driven printer arrived in 1979. Personal computers have taken over both the prediction and production of the tables since 1987. Today, the official publication of the Tide and Tidal Current Prediction Tables is no longer a book. The entire seven volumes fit on one CD-ROM.

In chapters two through four, we learned how the orbits of the earth, moon, and sun move the oceans by gravitation and momentum, while the rotation of the earth sends the tidal wave rushing toward the continents. In chapter five, we came to understand that the global tidal wave is actually the summation of many constituent waves, derived from many celestial cycles. In chapters six through nine, we learned how these waves are diverted by the continents and the Coriolis force, and then pushed up higher at the continental shelf. Continuing through these chapters, we learned that between the continental shelf and the shoreline these waves are further amplified by the topography of the near shore bottom and the contour of the coastline. Finally, within some coastal bays, resonant co-oscillation and sustained forcing of tidal energy may produce even higher tides. Coastal Kelvin waves may enhance the tide within channels near the coast. A seiche may complicate the tide pattern when waves from the previous tide interact with the following tide, or an estuary may distort the flood and ebb. Thus, the local beaches, bays, and harbors compose the oceanic tidal wave into unique local patterns, just as different musical instruments compose wind into music.

Now the ocean, straining toward the moon and sun, has collided with a continent and the tidal wave has broken against the shore. Just as a rolling ocean swell rises up at the beach and finally splashes into foam and spray, our long smooth gravity wave has been amplified and then fractured into variety by the immovable and irregular coast. Now all we have to do is reassemble this turbulence into a simple and reliable tide table.

(10-3)   Calculation of tide tables

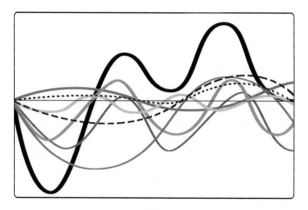

Reinforcement and interference of dozens of tide constituents summate into tide pattern (<u>black line</u>)

(10-4)

- Rotation of earth, as seen from moon
- Rotation of earth, as seen from sun
- Declination of moon's orbit to equator
- Declination of sun's orbit to equator
- - - Elliptical orbit of moon around earth
- Elliptical orbit of earth around sun
...... Resonance of lunar tide in shallows
— Deformity of oceanic waveform in shallows

(Frequencies are not to scale — for illustration only)

Illustration 10-3 provides a basis to begin our discussion of tide table calculation. Notice that multiple constituents of the tide are superimposed together on a graph. The dark black line is the predicted height of the water, the tide table. Each of the smaller, colored lines represents a tidal constituent that has been discussed in the previous chapters.[a]

When you add up all of the constituent curves (compute all of the reinforcement and interference between them), the sum total is the dark black line — the tide table for one specific location on the coast. Illustration 10-4 provides a partial list of these constituents. Using only this partial list, you could produce a tide table with about 90 per cent accuracy. All 37 of the constituents currently used by NOAA to compute their tide tables will be found in chapter thirteen.

Before we begin to calculate our tide table, we should consider the nature of these constituent graph curves, the smaller, colored lines in illustration 10-3. **Each constituent graph has three main characteristics: frequency** (waves per day), **phase** (the delay between passage of the sun or moon over a particular tide monitoring station and the crest of the tide graph for that constituent — see "establishment" on page 89), **and amplitude** (height of the water due to that constituent at that location).

---

[a] The constituents of the tide tables are almost entirely based on astronomical cycles. A few earthbound, non-celestial forces, such as the shorter frequency overtides and tidal pumping can be expressed mathematically as cyclic waveforms. These are included in the list of tide constituents. The only aspect of the weather that is accounted for in the constituents of the tide tables is the seasonal heating of the oceans, which expands the water and raises the sea level in a predictable pattern. Some local variables such as the contour of the coastline and the topography the near shore bottom are very important determinants of the height of the water, but they are not programmed into the computer as tidal constituents. Instead, their influence on the harmonic constituents at each tide monitoring station is determined empirically by measuring the previous year's tides. It is then assumed that they will have the same influence the following year. This assumption only fails when the physical parameters of the coastal basin are altered.

The *frequency* of each cycle is determined by the astronomical orbits of the earth, moon, and sun. The *phase* (establishment) of each constituent is determined by the local parameters over the continental shelf and at the shoreline and within the backwaters at each tide station. The *amplitude* is unique at each station because the small vertical displacement of the ocean is amplified by the local influences at that point on the coast.

These local geographic and hydraulic forces have been discussed in previous chapters. They include: (1) the shape of the coastline, (2) the topography of the near shore bottom, (3) the seiche effect that may elevate the water by a standing wave over the continental shelf or within a basin, (4) the amount of co-oscillation at the junction between the ocean and a coastal basin, which occurs when the period of natural oscillation at the coast is attuned to the period of the tidal wave (resonance), (5) the sustained forcing of tidal energy into such a coastal basin by successive tidal waves, (6) the amount of tidal pumping in that location, (7) the presence of coastal Kelvin waves, and (8) the distance to the nearest amphidrome.

Now that we are aware of the numerous earthbound influences on the tides, we will revisit a central fact stated in chapter four, and see how it can be used to calculate a tide table. **Variations in the height of the water and the timing of the tide cycles seen on different dates at the same coastal location are due to astronomical influences: the orbits of the earth, moon, and sun.** These cycles repeat every 18.6 years, and we already know this entire series of astronomical variations. By measuring the constituents for the past year, we can learn where we are in the 18.6 year cycle. Projecting forward in the known cycle tells us what the astronomical influences will be during the following year. **Variations in the height of the water and the**

timing of the tide cycles seen on the same date at different coastal locations are due to earthbound influences: oceanography, hydraulics, and geography. These are dependent on geophysical parameters, which, for practical purposes, have the same influence on the tide year after year. Therefore, by measuring how these local parameters influenced each constituent during the previous year, we can predict their influence during the coming year.

(10-5)

Step 1. Measure one year of tide cycles.

Step 2. Mathematically extract the constituents.

Step 3. Correct the harmonic constants which caused any discrepancies last year.

Step 4. Project the constituent curves into the coming year.
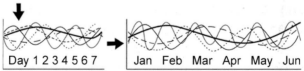

Step 5. Summate the constituents into the tide table.

Now we can begin to study the anatomy of a tide table. Anatomy can be properly learned only by dissection. Likewise, the only way to really understand a tide table is by taking it apart and putting it back together again. Illustration 10-5, above, depicts the five steps involved in the computation of a tide table.

Step (1) *The water level is recorded continuously during a series of successive tide cycles at a tide monitoring station*, and the hourly heights are tabulated. Theoretically, only 29 days of tide cycles (approximately one synodic month) are required to make a reasonably accurate prediction of the tide for the next year. In order to maximize the accuracy of the future tide table, it would be necessary to record every tide cycle for 18.6 years,[b] which is the period of time that includes every known variation of the celestial cycles. In practice, successive tides are measured for a period of twelve months, using a self-recording tide gauge.

The National Water Level Observation Network (NWLON) continuously operates 175 permanent water level stations. NWLON sends this data by GOES satellite transmission to the National Ocean Survey (NOS), which is an office within NOAA. In order to compute the corrections for coastal locations between these primary stations, NOS collects shorter term data from many other secondary monitoring stations.

At this first stage of the process you have a graph of the tide cycles at that location for the past year (the dark black line in Illustration 10-3).

---

[b] 18.6 years is the longest astronomical cycle with significant influence on the tides: the regression of the moons nodes. It is the time required for the maximal declination of the lunar orbit to cycle from 28.5 degrees from the equator, to 18 degrees, and back to 28.5 degrees maximal declination.

(10-6) National water level observation network (NWLON) station. There are 175 such permanent stations in North America.

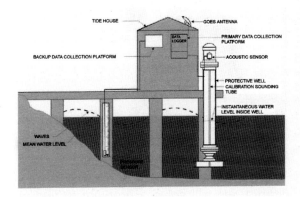

An acoustic sensor measures the water level. The sensor is encased in a cylinder, which excludes the wind-driven waves and surface motion.

Step (2) *Using a complex mathematical process known as "harmonic analysis," the independent contribution of 37 constituents can be extracted from the recorded data on the tide graph, for this specific point on the coast* (producing the multiple smaller colored lines in illustration 10-3). *Although it is not necessary to understand "harmonic analysis," we can demystify it, and even understand it in principle. By harmonic analysis, dozens of constituent curves can be extracted from a measured tide pattern. How is this possible?*

(a) By simple algebra we can find the value of an unknown, as in:

> (1) If (x) plus (y) equals 10,
> (2) and the value of (x) is 6,
> (3) what is the value of (y)?
> (4) We can solve for (y) equals 4.

(b) Next, recall that a curve on a graph is nothing more than a series of numbers, connected by a line. See Illustration 10-7(a).

(10-7) (a)   A graph is a spacial arrangement of numbers.

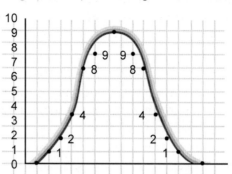

(c) Now, again using algebra, we will solve a problem to find an unknown curve on a graph, just as we solved for an unknown number. See illustration 10-7(b).

(10-7) (b)   If curve (------) plus curve (———) equal:

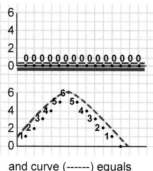

and curve (------) equals

This is the solution for curve (———):

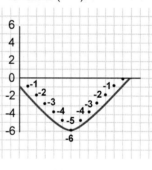

We can analyze graph 10-7(b) in the following way:

(1) If (curve x) plus (curve y) equals 0,0,0,0,0,0,0,0,0,0,0,0,0,
(2) and the values of (curve x) are 0,1,2,3,4,5,6,5,4,3,2,1,0,
(3) what are the values of (curve y)?
(4) Solving for (curve y) = 0,–1,–2,–3,–4,–5,–6,–5,–4,–3,–2,–1,0.

Of course, the mathematics required to extract multiple unknown curves from a measured tide pattern is much more advanced than simple algebra. The above example merely provides some insight into how such things are mathematically possible. Remember that any number of unknowns can be derived, if you can provide equations with enough knowns. Specifically, you can solve an equation for one unknown factor, provided you have at least two equations with that factor expressed in terms of other factors that are known. You can solve for two factors, if you have three such equations, and so on. With enough computer power (or the lifetime of a dedicated monk) you could solve for 114 unknowns, if you had at least 115 such equations. The *frequencies* of all of the constituent curves are known because they arise from the astronomical cycles. The *phase*, or establishment, for each constituent is known, as it will be the same as last year. In addition, the *sum* of all of the constituent curves for the previous year is also known — we measured it on a tide gauge. The only unknown is the future *amplitude* of each constituent curve at that tide monitoring station.

But we can gain a better understanding of harmonic analysis, which is at the heart of tide table calculation. Oceanographers around the globe are fond of studying their local coastal waters in order to learn the separate and distinct contribution of the lunar semidiurnal tide, and the contribution of the solar diurnal tide. You will recall our discussion of the basin in Thailand, where the semidiurnal tidal wave revolved around in the opposite direction as the diurnal tidal wave. In another example, the illustration below shows that around the entire globe there are actually two sets of amphidromes resulting from the Coriolis force, a pattern of amphidromes for the semidiurnal tidal wave, and another pattern for the diurnal tidal wave.

This begs the question: how do they know the distinct contribution of these two simultaneous waves to the height of the water? Actually, it is quite simple; and it can serve as a simplified model for how harmonic analysis extracts the constituent tidal curves from the measured tide pattern.

(10-8)

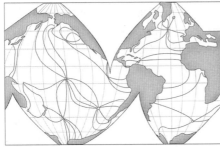

Semidiurnal amphidromes.
The dominant constituent is M2.

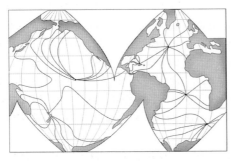

Diurnal amphidromes.
The dominant constituent is K1.

Two of the greatest celestial influences on the tide are the "principal *semidiurnal* lunar constituent," also known as M2, and the "lunisolar *diurnal* constituent," known as K1. This K1 constituent has a period of 24 hours (actually 23.93 hours), whereas the period of M2 is 12 hours, 26 minutes. If you simply measure the height of the water at noon every day for a series of successive tides, you have already isolated K1 from M2. This is because M2 doesn't repeat itself every 24 hours. It will be high "M2 tide" and low "M2 tide" an equal number of times over a long enough sample, and M2 will just cancel itself out for you, leaving only the K1 tidal wave contribution, at noon. Next, repeat the measurement at 1:00 p.m. every day, 2:00 p.m., 3:00 p.m., ...11:00 a.m. every day, and you will have isolated the entire K1 tidal wave from the M2 tidal wave. You will have extracted K1 from the entire summation tidal wave because K1 is the only constituent with a period of 24 hours, and you took your sample every 24 hours. This process is closely akin to the mathematical process of harmonic analysis. It is possible because: (1) you know the periods of M2 and K1, and (2) you know their sum, which is the same as the measured height of the water. You are using these "knowns" to find an "unknown" mathematical quantity, the height of the water due to the contribution of the K1 tidal wave.

Whereas it is relatively simple to measure the water level every 24 hours, and allow every constituent except K1 (every constituent with a period other than 24 hours) to average itself out of the tidal wave measurement; this would be rather tedious if the period of your constituent was 6 hours, 13 minutes, for instance, or every 13.66 days. In step one, above, the National Ocean Survey (NOS) has measured a continuous tracing of the height of the water for the past year. When they program a computer to follow this tracing, and select the measurement that occurred at that tide monitoring station every 6 hours, 13 minutes, they are asking the computer for the constituent curve of M4, the primary shallow water overtide of the semidiurnal tide (period of 6 hours, 13 minutes). When they program the computer for the water level measured every 13.66 days, the computer gives them the constituent known as O1, the lunar fortnightly effect due to lunar declination (period of 13.66 days). By this mathematical process,[c] you can extract each of the constituent tidal waves passing

---

[c] Mathematicians may want the following details, which are not at all essential to understanding tide table calculation: Operationally, for tide table calculation, the NOS uses a "least squares" harmonic analysis of 12 months observations of heights, that results in amplitudes and phases for 37 constituents. A backup Fourier analysis is sometimes used for additional analysis and quality control. This Fourier analysis uses 29 days of observed hourly heights to produce amplitudes and phases of 10 primary constituents directly. One synodic month of data is statistically inadequate to produce over 10 harmonic constants, and the NOS infers 15 additional secondary constituents from tidal theoretical relationships.

through each tide monitoring station from the measured height of the water during the previous year.

Each celestial cycle has a frequency just as a musical note has a pitch. The assembly of all of these celestial frequencies within the tidal wave is harmonic, just as the combination of simultaneous musical notes is harmonic. The computer model of each constituent is called a harmonic constant, and the mathematical process of extracting these constituents is called harmonic analysis. At this second stage of the process, you have 37 graphs of the constituent tidal waves passing through that tide monitoring station for the past year.[d]

Step (3) Every year, the previous tide predictions for each location are compared to the actual measured tide that occurred following publication of the tables. It is not uncommon to find discrepancies or anomalies: unexpected variances from the tide table predictions. When this occurs, *NOAA fine-tunes the computer program by updating those harmonic constants that are the source of these discrepancies*. We will discuss the reasons that some harmonic constants become obsolete and cause discrepancies from the tide table predictions, later in this chapter. At this stage of the tide table calculation process, by correcting these sources of error, you have an improved set of harmonic constants for each of the 37 constituents.

---

[d] In practice, not all 37 constituents are used for each tide station, if analysis shows that their contribution is negligible at that location. As a practical matter, NOS sometimes learns from years of experience with previous year's predictions that some of the 37 constituents are insignificant at certain NWLON stations. Furthermore some of the constituents are still reliable for that station for many years, they do not need to be measured every year, and they are reused until they begin to produce discrepancies with the measured tides. When this is the case, they use less than 37 harmonic constants for that location in the future. On the other hand, at some locations located up tidal rivers, up to 114 constituents are included. This is necessary to account for the complexity of a distorted tide curve due to the frictional estuary effect.

Step (4) *Each of the 37 constituent cycles can then be projected into the coming year, for that coastal location.* This is relatively simple, because we know the frequencies of the astronomical cycles that generate each of the constituent curves. In addition, we know the entire series of cyclical variations that repeat every 18.6 years. Once you know where you are in the cyclical process, you can just continue that cycle for one additional year. At this point you have 37 graphs of the constituent astronomical cycles for the coming year.

Step (5) *The 37 projected constituent curves are summed for the coming year.* In other words, **the reinforcement and interference of all 37 independent constituents is calculated for each minute of each day of the year. The tide table for each location is the sum of 37 graphs of constituent tidal waves that will pass through that location on the coast**. Now you have a tide table for the coming year, ready for distribution to the public, based on mathematical analysis of one year's recorded data.

There is one additional aspect of tide table calculation, which is restricted to certain locations where there is minimal tidal movement. Chesapeake Bay and Galveston Bay are examples of locations where the effect of the weather drives the water level just as much as the tidal forcing. Since the weather cannot be accounted for in tidal prediction by harmonic analysis, the tables have always been unreliable in such places. In recent years, the NOS has provided these ports with an additional tide prediction service. Every 12 hours the current meteorological data is programmed into a "numerical hydrodynamic nowcast/forecast" computer model. This provides the ship captains on these waters with real-time updates of the predicted tides for the next 12 hours: updates that include both the tide constituents and the weather.

In the introduction to this book we compared our ignorance to a boat that was grounded on a sandbar. We said that as we proceeded through these pages we would gradually accumulate facts, but it would be like the tide rising, inch by inch. We would have more water under us, but we would not be any more afloat than we were before. We said, "This is one of those complex subjects that require a gradual accumulation of facts until suddenly you are no longer aground." Each reader will have to judge for themselves, but I think that you have made that breakthrough when you look at a tide table, and:

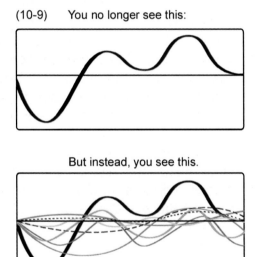

(10-9)   You no longer see this:

But instead, you see this.

I grew up in Florida and spent a lot of time at the coast. Most children in Florida have an intimate relationship with the beach, and I thought I had a better than average appreciation for the ocean. However, thirty years later when I began to scuba dive, I became familiar with the ocean floor, with its coral reefs and occasional shipwrecks, all covered with strange extra-terrestrial plants, and surrounded by an incredible variety of creatures, great and small, bizarre creatures that drift on the currents, and beautiful, perfectly adapted swimmers that spend their lives hoping to find something to eat, and trying to avoid being eaten by other beautiful, larger, ancient creatures. Now when I see the ocean from the beach, it looks completely different to me. I have learned that **the surface of the ocean is not the ocean.** In a similar way, we have looked beneath the surface of the tide tables, we have looked at the tides in depth, and I suspect they will never look quite the same.

In spite of computer science and our advanced state of astronomy, oceanography, mathematics, and physics, the tables are still based on empirical measurement of recent tides, and the computer calculations still require some degree of empirical correction. In other words, **in an age when we can land a man on the moon, we still cannot predict the tides by pure calculation**. There are a number of reasons for these unpredictable tides which cause discrepancies in the tide tables. They may be due to silting and sedimentation, with shoaling of the bottom, or even changes in the vegetation. Anomalous tides may be due to a change in the shoreline due to erosion, deep dredging by man, or manmade structures. Changes may arise from variation in the water temperature, or change in the viscosity of the water due to freshwater runoff. They also may be caused by something called "Chaos Theory."

In the past 25 years a new branch of science known as Chaos Theory has been developed. It seems to me that this theory is poorly named, because its main principle is that nature is *not* chaotic. Nature is deterministic, regulated by the laws of science, and controlled by known physical forces which can be described by mathematical formulas. However, nature is infinitely complex, because there are so many of these forces acting simultaneously. Although there are patterns in how they interact, they don't always behave according to those patterns, and sometimes they interact in a way that is totally unique. **Natural systems such as the weather and (to a much lesser extent) the tides are almost impossible to predict perfectly, because multiple independent variables can interact with each other in so many different combinations.** Incidentally, this is the reason that the stock market functions effectively and is impossible to outsmart.

This principle of "deterministic but unpredictable" pervades all branches of science. Consider the weather. We know that the formation of clouds depends on the air temperature, humidity, barometric pressure, and wind velocity and direction. We can reduce each of these forces to simple physics and mathematics, which is to say they are deterministic. Meteorologists can warn us that it will be cloudy tomorrow, which direction the clouds will come from, and what type of clouds to expect. But not even Einstein, himself, could predict the exact shape of an individual cloud, at a certain place in the sky, at a certain time.[e]

(10-10)

Every cloud is the result of a unique combination of:

Barometric pressure, air temperature, humidity, wind velocity and wind direction.

---

[e] The analogy between clouds and the tides is useful to introduce the concept of how nature is deterministic but somewhat unpredictable. However, the author does not mean to imply that the tides are as unpredictable as a cloud. The tides occur in predictable cycles which repeat in daily patterns that can be described as waveforms and analyzed by mathematics. The clouds are so complex that they defy analysis, and they do not repeat in any recognizable daily patterns.

The influence of Chaos Theory on the tides, the contribution of Chaos Theory to the occasional anomalous tides that conflict with the tide tables, may be somewhat controversial. Oceanographers differ in their opinion of the role of Chaos Theory in this context. I am not formally trained in oceanography. I only know that every other system in which there are hundreds of independent, nonlinear, simultaneously interacting constituents will have a degree of unpredictability. There is no way that the tides could be the only system in nature that does not behave in this way. **The tides are deterministic, governed by known forces, and very predictable. But they are not perfectly predictable.**

In her discussion of harmonic analysis of the tides, oceanographer, Dr. Sybil Parker, almost gave a definition of the effect of chaos theory, saying, "Since the length of the record is finite, the harmonic constants of the constituents are somewhat contaminated by the effects of the other constituents." We could almost, but not entirely, eliminate this unpredictability by: (1) using all 396 known tidal constituents in our computer model, instead of 37 constituents, (2) measuring the water level at tens of thousands of tide monitoring stations, located every mile along every coast, instead of a few hundred stations, and (3) measuring these tides for 18.6 years at each station, instead of one year.

Why would using the entire 18.6 year cycle of the near solar system "almost" eliminate uncertainty from our calculations? Why wouldn't this completely eliminate all unpredictability from the system? Because of the following principle, which is the mantra of chaos theorists: "All complex natural systems are highly dependent on initial conditions." This principle recognizes that whenever any scientist measures a natural system that is changing, he must begin his measurements at *some* time in the past (or the present). The problem is that something was happening *before* he began taking measurements. The system was also changing during the time leading up to the beginning of his study. Each time you go back further into the past, your conclusions become more and more accurate. But no matter where you start, your conclusions will come out differently, if you just start measuring the system at an earlier point in time.

Using all of the known mathematical formulas that represent all of the known cycles in the near solar system for 18.6 years will produce a very accurate tide table, but not a perfect one. This is because both the earth and the near solar system were dynamic and changing in the time preceding our first data point. The earth is obviously evolving over time, by such things as the shift of tectonic plates, climate change, etc. That is why the sea level is reevaluated every 25 years.

The near solar system is also changing, although the time frame is much longer. For example, the moon gets farther and farther away from the earth over eons of time, due to a transfer of energy between the tides, the rotating planet and the orbiting moon (page 87). Likewise, the rotation of the earth is gradually slowing. Then, there is the precession of the earth in its solar orbit, a gradual realignment of the earth to the fixed stars. Without doubt, there are many other such examples of changing parameters in the near solar system that affect the tides. **Because the "initial conditions" of any study change as you go back in time, infinitely; it is literally impossible to eliminate some degree of unpredictability in the scientific study of ever-changing nature.**

There is another consequence of Chaos Theory for all natural systems including the tides. Whereas complexity makes nature somewhat unpredictable, this same complexity makes nature more stable. What does that mean? Let's look at some examples of how complexity creates stability.

The electrocardiogram (EKG) is a graphic tracing of the electrical voltage produced by the heart, during contraction and relaxation. See illustration 10-11. In fact, this measurement of the heart muscle voltage, taken over the outside of the chest wall, is an average voltage produced by millions of individual microscopic heart muscle fibers. Each muscle fiber has its own miniscule voltage, and the EKG is the sum total of these cellular voltages. The EKG tracing in 10-11(a) indicates that the voltage of all of the heart muscle fibers is most positive at point (r) and most negative at points (q) and (s). The simplest system that could create this result would be for every single heart muscle fiber to be perfectly synchronous, as in 10-11(b). In fact, however, the muscle fibers do *not* contract in perfect synchrony. They are slightly asynchronous, as in 10-11(c).

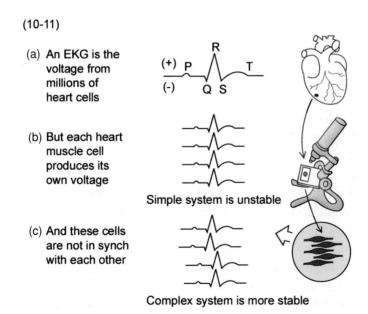

(10-11)

(a) An EKG is the voltage from millions of heart cells

(b) But each heart muscle cell produces its own voltage

Simple system is unstable

(c) And these cells are not in synch with each other

Complex system is more stable

Some of them reach point (r) a microsecond earlier than the others, and some of them reach point (r) a microsecond later (and some are 2 microseconds early, and some are 3 microseconds late, etc). This makes this system much more complex than one in which every fiber contracts simultaneously.

Whether you believe that this system was the result of evolution or a grand design, you must concede that it could have been a simpler system in which every heart muscle fiber beat in perfect synchrony. But, it's not. It is a very complex system with millions of muscle fibers beating slightly out of step with the others. It turns out that the system is more stable because of this complexity.

There are many risks that can lead to a cardiac death. In the end, all of them are attended by a fatal heart arrhythmia (an ineffective fluttering heart rhythm that cannot pump enough blood to sustain life). An electric shock or a mechanical blow to the chest is more likely to cause a fatal arrhythmia, if it is timed at the most vulnerable point in the EKG cycle of electrical activity. If every heart muscle fiber is at this most vulnerable point at the same time, every heart muscle fiber will be destabilized, if such a shock occurs during that microsecond. Because of its simplicity, one chance event could affect all of the heart muscle fibers at the same time.

On the other hand, if a shock or other destabilizing event occurs when some of the muscle fibers are *not* at the vulnerable point in the EKG cycle, there is a good chance that these "out of synch" muscle fibers will survive the event. They will maintain their normal electrical cycle and they may be able to "pace" the destabilized fibers back into a normal rhythm. Because the more complex system avoids all of the muscle fibers being affected by one chance event during the same microsecond, this **complexity creates stability**.

This concept is pervasive throughout the natural world. It occurs on a microscopic scale, as we have just seen; and it occurs over and over again on the larger scale that we inhabit. Let's consider the life cycle of one of nature's most amazing creatures, the Pacific salmon. See illustration 10-12(a).

(10-12)

(a) Entire species vulnerable to single natural disaster

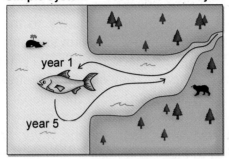

Simple system — 100% follow life cycle

(b) Complexity — unpredictable individuals and events leads to stability of entire system

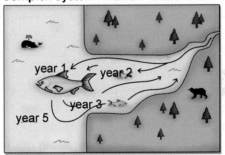

Complex system — small % late ocean migration

— small % early spawning migration

In the summer, the female salmon lays her eggs amongst the stones of a creek bed, the males fertilize these eggs, and then all of the adults promptly die.[f] This provides food for the bears, eagles, and other residents of the protein-poor tundra ecosystem.

---

[f] See the supplement to chapter ten, at the end of this chapter.

The following spring, the salmon frey hatch out in the upper reaches of these snowmelt streams. Those that survive predation by other fish species migrate down their native stream and out through a river into the ocean, where they feed for about five years on the protein-rich baitfish that proliferate in the cold, deep north pacific. On reaching their maximal size and sexual maturity, they make the long journey upstream to the place of their birth, and the cycle begins anew. It is not this simple however. A small percentage of the juvenile salmon do *not* leave their native stream during their first year. They stay in this stream and its attendant rivers for one extra year before they depart for the ocean. See illustration 10-12(b). They do this in spite of the fact that the food supply is relatively poor in the river, putting them at a disadvantage to their ocean-going siblings.

Furthermore, out of the majority that go to sea, a small percentage of Pacific salmon will *not* wait for five years before they return to their spawning stream. These nonconformists may go up to the spawning beds after only one or two years. Although they may be capable of reproduction, they are unlikely to find a mate, because they cannot compete with the larger males (the ritual of pairing off into spawning pairs involves fierce fighting among the males). The life cycle of the majority is simple and predictable. The exceptions to the life cycle, the nonconformists, create complexity in the system, and it is difficult to see any purpose for this complexity at first glance.

Suppose that there is a natural disaster in the ocean after the yearling salmon depart their birthplace stream. All of the adults may be killed by a single event. In this case, the few yearling salmon that stayed in the river for another year will still be available to quickly restock the species. If there is a natural disaster in a tundra ecosystem stream during one season of spawning, fish that come upstream early in their life cycle will begin to replenish the species immediately. It is the complexity of salmon behavior that provides stability for the species at times when the typical behavior makes them vulnerable to extinction by natural disasters.

At some point in their education, all science students are bound to wonder, "Why does everything have to be so complicated?" Well, now you know the answer — it doesn't have to be complex, but **it is complex, because complexity works better than simplicity. The same complexity that prevents the predictability of an individual or a specific event also provides overall stability of the entire system**. Why spend so much time on this subject that is peripheral to the tides on earth? Because, it is appropriate to recognize the essential role of complexity in nature, and examine the nature of complexity itself. After all, **if there is a central theme to the previous chapters, it is to emphasize the complexity of the tides: to create a new paradigm, with complexity replacing simplicity, by looking beyond the moon and lunar gravitation in order to understand the tides on earth**.

And so, it is not possible to go out on the ocean with an instrument and physically measure each constituent cycle within the tidal wave, because they are so numerous, and because they can occur in so many different combinations at different times. By the same token, because of the infinite complexity of nature, it is not possible to take the measurements of a basin and predict the pattern of the tide that will occur in that place by pure mathematics — that is, without measuring recent tide patterns (the first step in tide table calculation). What *is* possible, however, is to record the recent tides at a tide station, and then knowing the forces of nature that generate the constituent tide cycles, generate a tide table for that location for the coming year. **We can do this because we have now identified the individual constituents, the component "driving forces" of the tides; and these same constituents create every tide table that we encounter.**

Readers were promised that they could make sense of every table, and then enlighten their unfortunate, ignorant boating companions. The truth is that after you close these pages you will encounter unique tide patterns, and it will be difficult to be absolutely certain which forces of astronomy, oceanography, hydraulics, and geography play the dominant roles at that specific location. What we must rely on is that: (1) the tides are determined by many independent and simultaneous forces that we *do* understand, and (2) these many forces reinforce each other and interfere with each other in so many different combinations that it may be impossible to be absolutely certain of the importance of any one force or influence, in any specific location, at any one time. However, (3) we *can* devise a rational explanation for any tide, because we now understand those fundamental forces of nature that create all tide patterns. And most of all, (4) **We now understand that almost any tide pattern is possible in this complex system, just as we understand that a cloud can have any shape**.

When we see a unique cloud we aren't mystified by it, even though we don't absolutely know every interaction of nature that created it. **We accept it as a reasonable possibility**, because we know that it resulted from the interaction of air temperature, humidity, barometric pressure, and wind velocity. From now on, when we encounter an unusual or unfamiliar tide table, we must adopt the philosophy that we may not know for certain which of the many forces of nature dominate that tide pattern, but we are not confused by any tide table, and **we can construct a reasonable explanation for every tide pattern that we encounter, based on our knowledge of the constituent driving forces and coastal influences**.

Where there was confusion, now there is imperfect knowledge. That is certainly an improvement, and it will have to be good enough for now.

Supplement to chapter ten:

It is somewhat depressing when you see thousands of carcasses of large salmon in their spawning colors, lying on the banks of Alaskan streams in various stages of decomposition. When I shared this sentiment with my Alaskan guide, he gave me his interpretation of this seemingly odd fact of nature. He said that in woodland ecosystems near the artic circle there is not enough animal protein to support the large mammals like caribou, wolves, moose, and grizzly bears. Only a few hundred miles away, the cold north Pacific Ocean is teaming with enormous schools of protein-rich baitfish, and the like. He said that it was as if mother nature had found a way to run an extension cord down the streams and out into the ocean, in order to bring energy from where it was abundant to where it was lacking. The tiny salmon hatchlings go out to sea and feed on baitfish for five years, each storing ten to fifty pounds of protein and energy into their bodies. Then they bring this protein back upstream to the relatively barren tundra ecosystem, reproduce, and then release this needed energy.

A year or so after I heard this fanciful notion, I read that biologists have analyzed the meat and bones of sub-artic mammals and birds, to learn the relation between their diets and their environment. They found that the ratio of radioactive isotopes incorporated into these woodland animals is exactly the same as the unique ratio of isotopes found in the fishes of the deepwater north Pacific.

Now when I see salmon carcasses in spawning streams, I no longer see dead fish. Instead I find comfort in the purpose and economy of nature. This is an important revelation, since this is the same nature that we share, the nature that supports us too.

Chapter Eleven

The Weather and the Tides. Atmospheric Tides

"Whatever may be the progress of the sciences, never will observers who are trustworthy and careful of their reputations venture to foretell the weather."

> Notice in the *Times* of London, June 8, 1864,
> upon firing the first weatherman.

Your local weather is not taken into account in the computation of tide tables. Therefore, the mariner must understand the effects of meteorology on the height of the water, and adjust the tide predictions accordingly. If he fails to account for the barometric pressure, wind speed and direction, and importantly, the *duration* of that wind during the preceding 24–48 hours, he may find that the tide tables alone are not adequate for safe boating. The influence of the weather on the height of the water can be very significant; and, obviously, no one can anticipate the weather conditions at the time that the tide tables are printed. The tables may indicate safe conditions on your voyage, and the weather may change the depth enough to make your passage unsafe. It is important to understand how to account for the barometric pressure, and the wind speed, direction, and duration, so as not to be inconvenienced or endangered by these variations from the tables.

The effect of barometric pressure on the height of the water is quite simple. The atmosphere has weight, and this weight presses down on the surface of the oceans. The atmosphere, from outer space down to the surface of the earth, weighs as much as 33 feet depth of water. The average atmospheric pressure is equal to the weight of a column of mercury 29.92 inches high. The weather changes as areas of high atmospheric pressure and low atmospheric pressure move around the globe (see illustration 6-4). "Highs" are usually associated with fair weather, and "lows" generally bring foul weather. As the atmospheric pressure drops below 29.92, the sea level rises (higher than the tide tables predicted), because there is less pressure on the surface. As the pressure rises above 29.92, the sea level is pressed down (lower than the tide tables predicted).

The highest pressure ever recorded was 32.04 inches of mercury. This would lower the surface of the ocean by about 30 inches. The lowest pressure ever recorded was 25.52 inches, which would have raised the sea level by 57 inches. These are very significant variations from the tide tables, and indicate that the barometric pressure has a real effect, which is worth consideration. However, these were the most severe weather conditions in recorded history — no one in their right mind would have been out on the water intentionally.

As a practical matter, the daily fluctuation in barometric pressure rarely causes significant variation from the tide tables. A typical fluctuation in pressure associated with day-to-day weather change might be in the range of 0.075 inches of mercury (high of 29.99, or low of 29.84). This fluctuation would be associated with a variation from the tide table of only one inch of water. An approaching weather front typically involves about 0.75 inches of mercury (high of 30.67, or low of 29.17). This would cause a variation of about 10 inches in the height of the water compared to the tide table prediction, a very significant variation for boaters. However, the weather forecast would surely advise boaters to stay ashore under these conditions.

Those who are caught by a weather front, when they are unable to seek out safe harbor, should consult a weather channel to determine whether the pressure is high or low. Most storms are associated with low pressure, which would *raise* the water level and make stranding *less* likely. Many experienced mariners claim to have had personal experience with low water variation from the tide tables that occurred at times of foul weather. However, these anecdotal reports are due to the fact that low pressure is usually associated with strong winds. Often, it was the wind that caused the low water level that they erroneously attribute to the barometer. In other words, **low pressure itself raises the water level and makes boating safer, but the associated winds may create a hazard for boaters by lowering the water level.**

The greatest risk comes from high pressure, which lowers the water level and may expose your hull to bottom hazards. This risk is compounded because high pressure is often associated with non-threatening clear skies. **When boating in unfamiliar waters or shallow local waters, mariners should check the barometric pressure before pulling anchor.** If a captain learns of high pressure, he should subtract at least one foot (or better yet, calculate the exact variation) from the tide tables in order to pass safely over the bottom.

How do you calculate the variation from the tide tables due to the barometric pressure? You only need to know two facts: **(1) normal atmospheric pressure is 29.92 inches of mercury, and (2) mercury weighs 14 times as much as water.** Since the barometric pressure is reported in inches of mercury, you simply multiply the variance from normal pressure (29.92 inches) by 14 to get the variance in inches of water.

Example: Barometric pressure is 30.35. Low tide prediction is 5 feet, 6 inches. Pressure variance is 30.35 minus 29.92 = 0.43 inches of mercury. Sea level variance is 0.43 times 14 = 6 inches of water. Actual low tide is 5' 6" minus 6" = 5 feet.

An easy rule-of-thumb is: *one inch* **of change in barometric pressure will cause about** *one foot* **of change in the sea level (actually 14 inches).**[a]

Changing barometric pressure and moving weather fronts are often associated with fishing success or failure. There are a few pages at the end of this chapter providing a brief look at the relationship between barometric pressure and fishing.[b]

Along the coast, the wind has a very significant influence on the height of the water. It is important to consider the wind speed and direction in order to anticipate variations from the tide table. Naturally, winds coming from the sea push water higher at the coast, and winds from the land push water out to sea, lowering the height of the water at the coast. It is also obvious that the greater the wind speed, the greater the effect on the height of the water. There is another important consideration — how long the wind has been blowing in the same direction. Relatively mild winds that have been steady for the past 24 hours will have a greater effect on the height of the water than stronger winds of only six hours duration.

---

[a] Barometric pressure is the same thing as atmospheric pressure, or air pressure. Research scientists measure air pressure in pounds per square inch, or PSI. Television weathermen often use inches of mercury, or inHg. Newspapers and other sources often use millibars, or mb. Millibars, like PSI, is a direct measurement of air pressure, rather than a reference to a column of mercury. The conversion is: 29.92 inHg = 14.69 PSI = 1013.25 mb. One reason to use 29.92 inHg is that one inch change in inches of mercury causes about one foot change in water level, whereas this would require 0.43 PSI, or 27.08 mb (not very handy numbers).

[b] See supplement to chapter 11.

These are all accurate statements, but they are merely generalizations. An anonymous wise man once said, "All generalizations are false, including this one." I am sure that you are waiting for a simple formula to calculate, "If the wind is blowing (x) miles per hour for (y) hours, the height of the water will change by (z) feet." Unfortunately, this cannot be done.

If you think it through, the wind cannot change the water level unless it creates a gradient or slope in the surface of the water — there must be a gradient between the altered water level (where there is wind-effect) and the normal water level (where there is no wind-effect). For this reason, the wind has a negligible influence on the water level in deep water, off shore. This is because the wind cannot create a slope or gradient in the surface of the water in places where it can redistribute itself with the surrounding ocean. The influence of wind is only realized in shallow water and where a body of water is confined by a landmass.

When the water is pushed into shallow confined bays and backwaters, it cannot run off and return to the surrounding sea level, and so it is elevated. When the water is blown out of confined shallows, it cannot be immediately replaced, and it remains lower. **The shallower the water, and the more it is confined by a landmass, the greater the influence of the wind on the height of the water**. This is the reason that there is no simple formula to calculate the wind-effect on the water. Since each body of water has a unique size, shape, and depth, there cannot be one simple formula to calculate the wind-effect for all locations.

The duration of a steady wind is just as important as the wind speed. It is not uncommon for 10 to 20 mile-per-hour winds that persist for 12 hours to cause a dangerous variation from the tide tables, within a coastal basin. 10 to 20 mile-per-hour off shore winds lasting 24 to 36 hours might cause over 12 inches drop in the water level along the coast. Many vessels have been stranded by low water due to off shore winds. **Although it is difficult to predict, the wind-effect can be very significant and should always be considered when boating in shallow water, or near submerged hazards.**

Most boaters spend the majority of their time afloat plying the same waters near their home port. Experience with the winds and tides in your locale will enable you to estimate the wind-effect after a few weather events are recorded in your log. You may record that 20 mile-per-hour winds from the north, lasting 12 hours, resulted in a 12 inch drop in the water level in your local harbor. This same event might cause a 15 inch drop in another harbor, or a 3 inch drop elsewhere.

If you take the trouble to learn how the wind affects the water level at your homeport, this will allow you to estimate the variance from the tide tables for decades to come. Of course, you can also learn such sea lore from the "old salts" that inhabit every harbor. By keeping a log of weather events and listening to local experts, you can be as good as anyone else at predicting the variance from the tide table due to wind.

There is one more parameter of the wind-effect, which is not available from the weather services. You can find out the wind speed, direction, and duration. But, the unknown quantity is the "fetch" of the wind. The fetch is a measure of the length of water surface that the wind is forcing. Thus, a 20 mile wide band of coastal water may be effected by the wind, or this same wind may push against the surface of a 300 mile wide band of coastal water. **The greater the fetch, the more friction between wind and water, and the greater the wind-effect.** This is usually an unknown parameter — another reason that the wind-effect is somewhat unpredictable.

Like everything else related to the tides, the sea level is effected by more than one weather influence most of the time.

For example, during many weather events with changing barometric pressure, there will be high winds. Depending on their direction, the winds may lower the water level or raise it. The changing pressure may raise or lower the water level at the same time. Often the best you can do is simply be aware that **the tide tables are not as reliable when the weather is changing**. Then, you can be extra cautious when planning your passage over minimal acceptable depth for your vessel.

The third meteorological influence on the water level is the temperature. During the summer, the water is warmer and it expands. In the tropical and temperate latitudes, this can raise the sea level about 6 inches. That is equivalent to the influence of solar gravitation. The thermal effect is progressively less as you proceed toward the poles. This effect is so predictable that it is included as a constituent in the computation of the tide tables. Therefore, mariners do not need to account for this seasonal influence.

Finally, recent rainfall may cause variation from the tide tables for inshore waters: coastal rivers and the intracoastal waterway, and its tributaries, and countless unnamed creeks. For several days following a heavy rainfall, run-off from the land may cause a measurable elevation of both the high tide level and the low tide level, inshore. On the other hand, a prolonged draught may cause saltwater incursion, but no significant variance in the height of coastal rivers and backwaters, which remain filled by the tides from the great ocean reservoir.

Early in my research into this subject, I was sitting in the beautiful public library in Camden, Maine, looking through a large volume on meteorology, when I came across a chapter that was all about the tides. I began taking notes, but soon found that the text was at odds with my other reference sources. Then, I realized that this chapter was about the *tides in the atmosphere*, not the tides in the ocean. I was already aware that the solar and lunar gravitation could lift the oceans a few feet, and even lift the continents a few inches. It had not occurred to me that the atmosphere of lighter material would also have a tide. The portion of the atmosphere where weather occurs is only about 24 miles thick. The atmospheric tides lift this layer about 1.1 miles — it would be surprising if this did not have a significant effect on the weather.

Lunar and solar gravitation both affect the barometric pressure by lifting the atmosphere away from the earth's surface. The lunar effect occurs in a semidiurnal cycle, and is greatest near the equator. The solar effect on barometric pressure is 15 times greater than the lunar induced fluctuation, because the solar influence is a combination of gravitation and radiation (heat). Warming the atmosphere causes expansion, a decrease in density of the air, and lower barometric pressure.[c] This solar influence has both a semidiurnal cycle and a diurnal cycle,[d] and is called "thermotidal oscillation."

The important result of lunar gravitation and the thermotidal oscillation is the creation of atmospheric tidal winds. Just as the weather-related centers of high and low pressure cause winds as air moves toward nearby "lows," the tidal changes in barometric pressure also result in winds. Near the surface of the earth, these tidal winds rarely exceed two miles per hour. However, at higher altitude the air is less dense, and the tidal winds are much greater. At six miles above the earth, these winds reach 20 miles per hour.

---

[c] A gas confined in a closed chamber increases pressure as it is heated. However, gas that is free to expand becomes less dense and decreases pressure.

[d] The diurnal cycle of thermotidal oscillation is the result of very complex thermal dynamics. This involves (a) the time required for the periodic heating and cooling of air, and (b) the vertical movement of heated air. Neither the temperature change nor the vertical movement can occur instantaneously. These dynamics add an additional diurnal cycle to the solar atmospheric tides.

At 30 miles up, they may exceed 100 miles per hour. Therefore the atmospheric tides are a very important aspect of meteorology in the upper atmosphere.

There are many examples of the most severe weather events in history that happened during a conjunction of maximal celestial influences on the tides: the tides in the ocean *and* the tides in the atmosphere. Because there are so many variables involved in the causation of the weather, it is difficult to prove that the tides in the atmosphere have caused storms or other weather events. However, there is plenty of evidence that the same gravitational forces that cause the ocean tides influence the weather by means of atmospheric tides. This is a field of science that needs further investigation. In the meantime, it is very likely that there is a linkage between the weather and the gravitation of the moon and sun.

Supplement to chapter 11:

Although there is little scientific data on the subject, all fishermen seem to have a personal theory on changes in barometric pressure, fish feeding behavior, and successful fishing. Popular fishing lore holds that: (1) fishing is more productive just before a weather front arrives (with changing barometric pressure), and (2) fishing is usually slow just after a weather front passes through. (3) Fish feed better when the pressure is rising than when the pressure is falling. (4) Fishing is relatively poor when the barometric pressure is very high or very low.

I respect the fact that many professional fishing guides and many sportsmen with a lifetime of experience have all reached these same conclusions independently, and they are absolutely convinced of a linkage between fishing success and barometric pressure. I too have cancelled out-of-town fishing trips that were scheduled for the day after a weather front passed over.

However, the following scientific facts should convince any reasonable person that a fish could not possibly be aware of the barometric pressure. Since the weight of the atmosphere is equal to the weight of 33 feet of water, when a scuba diver is 33 feet below the surface, he will experience "two atmospheres" of pressure on his body — one atmosphere of pressure from the weight of the *atmosphere* above him — one atmosphere of pressure from the weight of the *water* above him. This is also true for fish. Even though they may have superior pressure sensory organs, and I am certain that they do; they must somehow deal with the weight of the atmosphere (barometric pressure) plus the weight of the water above them. Now, suppose there is a weather system approaching and the barometric pressure falls 0.75 inches of mercury. This might be detected by a fish, *unless* the fish changed depth in the water column. As soon as the fish moved down 10 inches deeper, the pressure on his sensory organs would be the same as before. All fish are moving and changing depth most of the time. Even if they held still, the pressure would change when they were below the crest of each surface wave compared to the pressure below the trough of the wave. Because water is so much heavier than air, the slightest change in the depth of the fish would outweigh the changes in the barometric pressure.

Anything is theoretically possible, but I cannot believe that the fish can somehow measure the distance to the surface second-by-second, and compute the correction factor necessary to arrive at atmospheric pressure. We have to form opinions about many things without absolute proof. We do it all the time. Juries of reasonable people condemn defendants to the gas chamber, without absolute proof. **Because we rarely have absolute proof of anything, we usually have to decide what is reasonable**. It is not reasonable that fish can know the barometric pressure. And yet, experts insist that weather fronts affect fishing.

How can we resolve this conflicting information? My answer is that roosters do not cause the sun to rise. I can't prove this; but that's my opinion. Throughout recorded history, everyone agrees that roosters crow and then the sun rises over the horizon. It is extremely reliable — soon after the rooster crows, the sun rises — it happens every time. The erroneous assumption of cause and effect is the source of a great deal of faulty logic and most junk science.

Could it be that changes in barometric pressure are associated with some other influence on fish feeding behavior? Approaching weather fronts are usually associated with cloudy skies, and low light levels. Skies are usually clear after a front passes by. Doubtless, there are many other changing parameters in nature associated with these dramatic shifts in the atmosphere. I suspect that there is some other natural occurrence associated with pressure change that creates a correlation between weather and fishing success.

Besides the implications for saltwater fishing, there is a moral to this story. (1) Beware of assigning cause and effect where it does not exist, and (2) do not hesitate to use reliable empirical information that you do not understand. I am convinced that fish cannot be aware of the barometric pressure, but the next time I am scheduled for a fishing trip right after a weather front passes through, I am going to cancel.

When penicillin was discovered, no one had any idea how it worked for almost twenty years. In the meantime, it saved millions of lives. During the decades before we learned the mechanism by which it kills bacteria, we did not hesitate to use it, because it works.

Chapter Twelve

The Tides and Saltwater Fishing

"So long as rivers run free and the tides roll, so long as fish press their fins against the waters, so long as the weather is changeable and man is fallible, just so long will the fishing be unpredictable, and joyful."   Olde Irish Ode

One of my fishing buddies is a retired veterinarian who never travels anywhere without his canoe and a compliment of rods and reels. Ed and I were flyfishing for smallmouth bass, last summer, on a beautiful stretch of the Penobscot River in Maine. Right after a lunch on the shore of an island, Ed shoved off and paddled straight for some inviting subsurface boulders, perfect smallmouth bass habitat. As soon as he dropped anchor, he caught and released four monster mossybacks (old bass). After he rejoined me where I was casting in less productive waters, I complimented him with, "Looks like it's your day, Ed."

"It sure is," he came back. "That's what I always tell my wife. Every now and then it's your day, and you never know for sure when it's going to come. So, you always need to be out on the water so you won't miss it when your day comes. I mean, you might be sitting around at home or working in the yard, and it would be your day, and you wouldn't even know it." These are words to live by, and it must be true; after all, Ed has a graduate degree from Auburn University.

When teaching fishing to youth groups, I always begin with this question: If you asked a fish, "Did you have a good day today?" and the fish responded, "Yes, I did," what would he mean exactly. In other words, what is a good day for a fish? The answer to this question contains the key to successful angling, because it tells you where the fish are most likely to be found. Here is the answer: **when a fish says that he had a good day, he means, "I found something to eat today, and nothing ate me today." The two most constant and essential needs of these creatures are to find food and avoid predators.** The oldest and largest ones are the ones you hope to catch. These are the ones that have learned to stay where they can find both food and relative safety from predators.

To be successful, the angler must also be aware of one primary difference in the behavior of freshwater fish and saltwater fish. **As a general rule, mature freshwater fish tend to be territorial, whereas saltwater fish tend to be constantly moving about in search of food.** To be more specific, the bass in ponds and streams will usually be hiding out near some sort of structure on the bottom, such as a fallen log, or a boulder. This is the best place to ambush smaller, dumber residents. In trout streams, the biggest fish will take up the best "lies" — places where the current is interrupted, and the senior citizen can rest in a pocket of quiet water, and let the food drift by like a moving cafeteria line.

In saltwater, along the coast, things are different. The tides dominate the world of coastal fish. **There is an intimate relationship between the tide cycles and the behavior of all saltwater marine creatures**. In an interesting experiment, a behavioral biologist relocated a hundred fiddler crabs, and established them in an artificially lighted laboratory. These crabs normally inhabit the tidal zone of a salt marsh. In the wild, they have a very regular pattern of emerging from their holes in the mud, and feeding in the spartina grass. The question was whether this pattern depended on: (1) the light level, (2) the moisture of the mud as the tide flooded the marsh, or (3) some other direct link to the tide cycle. In the laboratory, the fiddler crabs maintained a regular pattern of living alternately below the mud or feeding above ground, and this was not influenced by the light nor the moisture of their laboratory world. For about two weeks after they were relocated, their behavior was totally linked to the cycles of the tide *back at their original home*. It was obvious that they had the specific tide table for their coastline programmed into their nervous system. Only the phases of the tide back at their original home determined their whereabouts and their feeding behavior. After a few weeks, they lost touch with this link to the tide cycle, and the scientists could manipulate their behavior with artificial light and moisture.

Periwinkles are snails that live in the tidal zone at the edge of the sea. One species, the rough periwinkle, has adapted to the tide cycle by developing gills that can take oxygen from the air as well as water. This allows it to live above the waterline most of the time, and be independent from the sea for reproduction and other activities. It is only submerged for a few days every two weeks, during the fortnightly spring tides. On these days it cannot lie on the rocks undisturbed, but is required to stay active in order to deal with the surging waters. When these creatures are taken a great distance away, to a laboratory located where the tide cycle is completely different from their home, they somehow "remember" the monthly tide cycles of their native shore. For several months, not only are they most active in their laboratory aquarium on the days of the spring tides (back at their home), their tissues change water content in a cycle that would allow them to survive exposure to the air on all other days.

Some mollusks only reproduce during full moons. Others always produce larva at quarter moons, but never at full moons. Looking closer at these different species, you will see that the larva of full moon egg-layers are dispersed by the tidal currents as part of their natural life cycle. However, the larva of the quarter moon egg-layers have a life cycle that requires it to cling to the leaves of seafloor vegetation — they don't want to be dispersed by the currents. Each species links their reproduction to a lunar phase that will assist them in their particular niche: full moons and strong currents disperse one species, quarter moons and weak currents enable another species to cling to local plants.

These few examples are not rarities of nature. All marine animals are intimately wedded to the lunar phases, the tidal flood and ebb, and the tidal currents. In just these few examples, we see how the tides influence the reproduction, the level of activity, the location, the feeding behavior, and even the physical make-up of sea creatures. As Rachel Carson says, "The pulse of life is adjusted to the rhythm of the tides."

Returning to saltwater fishing, **rule number one is, "In order to find fish, you must find the natural bait, and the natural bait moves with the tidal currents."** Rule number two is, **"Memorize rule number one."**

The tides force the currents along the coast. These are the currents found near shore and in the rivers, bays, and estuaries that make up the backwaters, and the tidal zone (this chapter does not apply to offshore fishing). These currents are too strong for most of the creatures that qualify as *prey*: shrimp, juvenile crustaceans, and baitfish. They cannot resist the strongest currents, and they are swept wherever these currents flow. The *predators* can go wherever they want, and they learn to position themselves where the currents provide the prey. The problem for them is that the tide is constantly changing and so are the currents. Therefore, saltwater predators are almost constantly moving about during each day, in order to find their food, which is being moved about and concentrated by the currents.

If you are a coastal boater, you are aware that there are all sorts of currents, from gentle drifts to raging rips. Big gamefish love the rips that disorient the baitfish and render them helpless. Sometimes the speed of the current is related to a physical constriction of the waterway, where a large volume is forced through a narrow channel. Usually, however, the speed of the current is due to either the range of the tide (how much water moves during each six hour flood or ebb), or the phase of the tide (which hour you are dealing with).

When planning a saltwater fishing day, your first concern is the weather, and your next concern is the range of the tide and the time of the high and low tides at your destination. If the range is large, more water will move throughout the tide cycle and the currents will be stronger. Along the coast, **fishing is usually best when there is a large tidal range**.

The rate of movement between the high and low tides is not linear. The water level changes more during some hours than others. See illustration 12-1. The "Rule of Twelve" is a very useful tool for the saltwater fisherman and all mariners.[a] Notice that the curve of the tidal movement is steeper at the point halfway into the flood tide.

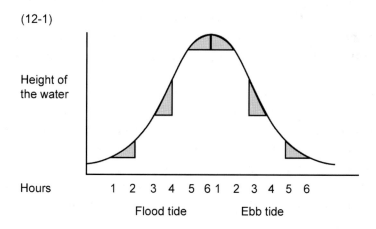

(12-1)

Height of the water

Hours   1 2 3 4 5 6 1 2 3 4 5 6

Flood tide   Ebb tide

---

[a] Throughout this book, the author has made a rigorous attempt to be scientifically accurate. The Rule of Twelve is not scientific and it is not rigorously accurate. It is a guideline, or "rule of thumb," but it is close enough to be very useful.

The water level changes 1/12th the first hour, 2/12ths the second hour, 3/12ths the third and fourth hours, 2/12ths the fifth hour, and 1/12th the sixth hour. Thus, if the range of the ebb tide is 6 feet, during the ebb tide the water will drop only 6 inches (1/12th of 6 feet) during hours one and six, the water will drop 12 inches (2/12th of 6 feet) during hours two and five, and, the water will drop 18 inches (3/12th of 6 feet) during hours three and four. The pattern is the same during the hours of the flood tide.

The currents are greatest during the third and fourth hours of each six hour tide cycle. Therefore, the concentration of the natural bait by the tidal currents is greatest during these hours, and least during hours one and six. Some species are more likely to be affected by this pattern. While fishing for Atlantic spotted seatrout (Cynoscion nebulosus), I will usually just stop fishing and have lunch or take some other break during the slack tide hours.

When fishing in very shallow water, there is another important consideration related to the coastal tides: the changing depth of the water. My favorite gamefish species, the "redfish" (Sciaenops ocellatus), has a preference for feeding in shallow water. This is an opportunity for this predator to find abundant baitfish and small crabs on the mudflats or other shallows, and at the same time avoid their chief predators, dolphin and sharks, which cannot follow them there (food and security, again). In shallow water, The Rule of Twelve is especially helpful to assure that you are in the right place at the right time.

Suppose that you find your targeted gamefish in 15 inches of water at a certain location on Saturday, January 1st, at 10:00 a.m. You record the place, time, and depth in your log. Two weeks later, you are planning a return trip to the same spot. *What you want to do is return there when the depth is 15 inches again, in that very location.* Here is your procedure. First, looking at the tide table for January 1st, you find that the water was 15 inches deep over your shallow water spot, 2 hours after low tide, at 10:00 a.m. Next, looking at the tide table for January 15th, you should use The Rule of Twelve and figure out what time of day the water will be 15 inches again at your destination. Here is how to make that calculation:

On January 1st, the low tide mark was 6 inches above MLLW. The tidal range was 6 feet. According to The Rule of Twelve, the water will rise 6 inches during hour one, and 12 inches during hour two. Therefore:

    (a) at low tide, the water level was 6 inches above MLLW.

    (b) after hour one, the water level was 12 inches above MLLW.

    (c) after hour two, the water level was 24 inches above MLLW.

(It is 15 inches deep at your destination — you measured it yourself. Note that the measured water level at your shallow destination is 9 inches less than the deep water level in the tide tables.)

On January 15th, the low tide mark is minus 6 inches. The tidal range is 8 feet. Now we start adding water from a lower point, and the water rises 8 inches during hour one, and 16 inches during hour two. Therefore:

    (a) At low tide, the water level is minus 6 inches below MLLW.

    (b) After hour one, the water level is 2 inches above MLLW.

    (c) After hour two, the level is 18 inches above MLLW.

(It will be 9 inches deep at your destination. When you measured it, yourself, on your first trip, the depth here was 9 inches less than the tide table level for deep water.)

In order to have 15 inches at your fishing "hotspot," you will need to return on January 15th at between 2½ and 3 hours after low tide. If you returned there two hours after low tide (because this was the tide phase on January 1st), you probably couldn't even float your boat into this location (only 9 inches deep on this trip). **Using the time elapsed after low tide (or high tide) will not give you the desired depth at your destination, because (1) the flood (or the ebb) does not always begin at the same water level on different days of the month, and (2) on different days, the water does not move the same amount each hour.**

Why not just return at 10:00 a.m. like you did on January 1st? This would probably be even worse. If low tide was at 8:00 a.m. on the 1st, it would be about 6:30 a.m. on the 15th. Therefore:

(a) At 6:30, the water level would be minus 6 inches below MLLW.

(b) At 7:30, the water level would be 2 inches above MLLW.

(c) At 8:30, it would be 18 inches above MLLW.

(d) At 9:30, it would be 42 inches, and at 10:00, it would be 54 inches.

Returning at the same time of day will not give you the desired depth at your destination, because the flood (or ebb) do not always begin at the same time on different days of the

month. **Only by using The Rule of Twelve, to calculate the hour-by-hour change in water level, can you guarantee the same depth at the same location on different days.**

This same principle should be applied to other fishing situations that depend on the depth for success. Perhaps you want to fish around an oyster bed when it is just barely covered by water. If you arrive too early, the fish will not be there yet; if you are too late, you probably can't even find the submerged structure. On a bonefish flat, the water first fills the channels around the flat, then covers the fishing ground, and then moves up into the mangroves on the shore. If you are too early in the flood, the fish will still be in deep water. If you are too late, the fish will be inaccessible among the mangrove roots, where they can find food in a place that provides protection from predators (predators like you). In all such situations, **the most accurate way to predict the changing water level is to access the change during each hour of the 6 hour tide cycle, rather than the time elapsed since the previous slack tide, or the time of day**.

Recently, two of my friends took a boat out fishing offshore. As luck would have it, they found a lot of fish just at sundown. Of course, they stayed too late. Returning to the marina in the dark, they turned up the wrong creek and ran aground on a falling tide. It was nine long hours before they could float out again. When the mosquitoes arrived, they had no bug spray. In order to survive the mosquitoes, they smeared themselves with motor oil. In order to survive the motor oil, they began to drink beer. They told me that in the middle of the night, the water receded so far from the boat that they couldn't even throw an empty beer can to the water's edge (they were in no mood to consider the ecology).

The tides can help you find the fish, but they can also cause hazards or leave you stranded on the bottom. This is where The Rule of Twelve can come in handy again. Suppose the tidal range is 6 feet. You are catching fish, but there are only 4 inches of water under your keel. The question is: do you have to "leave them biting?" The answer is: that depends on which hour of the tide cycle you are in. If it is hour three or four, the water will drop 4½ inches in the next 15 minutes, and you had better pull your anchor. However if it is hour one or six, the water will only drop 1 1/2 inches in the next 15 minutes, and you can stay a little longer. This alone makes it worthwhile to use this rule until it becomes second nature.

I belong to a large club for saltwater flyfishermen. We make an effort to provide education for the new members. One of the most common questions that I am asked is, "What is the best tide to catch fish, incoming or outgoing?" Unfortunately, there is no short answer to this question. The complexity of nature is truly mindboggling. There is no doubt that fish develop rather specific patterns of feeding behavior. However, each species has different patterns, and the same species will have different patterns in different locations, in different creeks, and in different parts of the same creek. For example, seatrout usually come into a saltmarsh creek with the incoming tide, and then move farther and farther up the creek as the floodtide progresses upstream. Then, they gradually leave the creek with the outgoing tidal current. This is really of little help, since it only tells you what happens in general during a twelve hour cycle.

What you want to know is what time will the school of seatrout pass by a certain dock, or feed near a certain oyster bed, or swim along the outside edge of a certain turn in the creek. This *is* possible, because **a school of fish will usually develop a regular pattern of feeding in each creek**. In creeks and deep waterways, you can time the arrival of the moving school by noting the time elapsed since the previous slack tide. For example, they may pass by a certain dock on the incoming tide, about two hours after low tide. **You can return there on different days, and have a good chance that they will repeat this pattern**. However, if you go to a different creek a mile away, the seatrout may routinely pass by another dock on the outgoing tide, about one hour after high, for example.

When I was young, I was fortunate enough to fish with an old salt who had 50 years of experience fishing in his local waters. He would motor some distance to get to a particular location, and then as we approached it, he might head off for another spot without casting even one time. As he turned the boat away, he might say, "They just won't be here unless the water is running north," or some such "local knowledge." At the next spot he would be looking for a completely different pattern of current, or a different phase of the tide. This was my first fishing lesson, and I never got a better one. **In tidal backwaters, the currents concentrate the natural bait, and create opportunities to locate feeding fish. Since the tides and currents are cyclical, the fish develop regular routines in their daily movement. This can be predicted for a specific location on the chart by noting the time elapsed since the previous slack tide. Because each creek and waterway has unique dimensions and contour, the currents are altered at each location. For this reason, the pattern of fish behavior is local behavior. Once learned, this "local knowledge" is the angler's most valuable asset.**

There is no alternative except to acquire your own local knowledge by experience, or by getting tips from fishermen who are willing to share information. In the meantime you return to the first principal and increase your odds by fishing where there is current (food) and structure (protection). This is what you must do when you travel to unfamiliar waters as well.

There is one other way to speed the process of becoming on old salt yourself. I once heard an interview with a twentieth century gold prospector. When asked, "How do you decide where to dig," he replied, "Where there's gold, there may be more gold." You can accumulate experience much faster by prospecting upstream and downstream from places where you have chanced upon a local fish feeding pattern.

It is very likely that a school of fish is moving in the direction of the current. Earlier in the tide cycle, they should be upstream at another oyster bar, grass bed, or dock before they reach your known hotspot on that same shoreline. Later in the tide cycle, they should be downstream at another feeding station, after they move on from your spot. **Always try upstream (earlier in the tide) and downstream (later in the tide) from good fishing, since saltwater predators are often moving with the tidal current that moves the bait.**

The other most common question that I hear from novice anglers is, "Do you prefer full moons, or which phase of the moon is best for fishing?" This can be answered with more authority than the incoming–outgoing question. The range of the tide is greater during full moons and new moons, and least during quarter moons (see chapter four). Large tidal ranges create strong currents and improved fishing. As a general rule, **new moons and full moons will be better fishing than quarter moons.**

There is another consideration however. During full moons there is more light for the predators to feed at night. Many saltwater species prefer to feed at night, and if it is cloudless during a full moon, the fishing may be slow on the following day. Like everything else in this sport, this is an inexact business, and fishing may be excellent during both new moons and full moons.

Why are some fishermen so much more successful than others? It is not because they have better boats and better tackle. It is not because they can cast better, or choose better lures. It is because they fish in the right place at the right time.

When I first started going out with more experienced anglers, I was impressed by their ability to consistently find feeding fish. We would be running down the intracoastal waterway, and suddenly he would pull in along the shoreline, saying, "This should be a good spot." As I scanned our surroundings, all of the water looked the same to me, and I would wonder how he knew that this location was any better than the rest. He would usually be right and the fish would be there more often than not, as if he could see them below the water like an osprey, or somehow sense their presence like a shark.

Years later, I decided that there are two kinds of fishermen. A very rare minority are truly instinctive fishermen. They are always from a rural background, and "grew up on the water." They have never read a chapter such as this one, and they don't need to. I suspect that either: (1) their grandmother was a Native American, or (2) their great grandfather once got lost on the Missouri River, and spent a long winter sharing a tent with a large otter, or something like that.

The rest of us (about 99.99 per cent of us) can learn to be productive anglers too. But only if we will focus our education on the most important aspect of fishing: the basic biological needs of fish that dictate where they will be feeding, and when they will be there.

The fish will be feeding where the natural bait is concentrated and disoriented by the tidal current. If they can do this where underwater structure provides an advantage over both their predators and their prey, that is better still. Besides learning how the biological needs of fish dictate where they will be feeding, there is one other critical aspect of successful fishing that we must all improve: *stealth*.

When you arrive at the right place at the right time, you must be stealthy. This is especially important in shallow water, but if you power up to a deep hole and toss out a heavy chain and anchor, the fish will regard you as a predator and relocate. Remember that you and I might get into the game on occasional weekends, but **every fish has been both a predator and a prey for every day of his whole life**. My formula for success is something like: 70% is location and timing, 20% is stealth, and 10% is everything else (boat, tackle, lures, casting ability, etc.).

The principle of using underwater structure to locate fish includes more than just fishing near oyster beds, grass, and docks. The bank of a creek may be vertical on one side (such as the outer bank of a sharp turn in the creek), and relatively shallow and more horizontal on the opposite bank. If a fish is in the middle of the creek, it is vulnerable to attack from *all* sides; if it is against the steep bank, it cannot be attacked from that side, and it is relatively safe. A fish in the middle third of the water column may be attacked from above or below; if it is on the bottom, it is safe from attack from below. If a fish can find a channel in the bottom or a shelf (abrupt vertical change) in the bottom, it is even less vulnerable from less directions.

Similar to the way fish use structure (docks, grass, oyster beds, etc.) to ambush prey, the light level is used by predators to gain an advantage. Remember that predators can go wherever they choose, whereas the small weak prey are pushed around by the currents, and even by the surface wind. If there is a light on a dock or bridge at night, there will usually be a shadow next to the lighted water. Recall how you can see into a lighted house if you are outside in the dark, but the people inside cannot see what is in the dark outside. Predators will hold a position within the shadow, and being unseen they can easily attack any prey that are swept into the light.

This is why fishing is best at dawn and at dusk, and on overcast days; the predators are out feeding when they can attack without being seen. In still water, such as ponds and lakes, this is one reason why fish prefer the shade and shadows, they feel less vulnerable.

Besides food and security, there are two other biological needs of all fish: tolerable temperature, and reproduction. Neither of these have anything to do with the tides. However, once you get an old fisherman talking, it is hard to make him stop. Let's briefly consider these other two biological needs, as part of a scientific approach to saltwater fishing.

We can quickly dispense with the significance of reproduction on fishing by saying that the biological need to reproduce only dictates the location of sexually mature fish (a small minority of the population), and only on a very limited number of days each year. Because of the variation in the length of days and the temperature, the month devoted to spawning by saltwater species varies from state to state along the coastline. Anglers must learn the spawning habits in their local fishery, in order to benefit from these patterns. Most of the time this will have no effect on the location of gamefish.

Water temperature is used by experienced anglers to locate fish. This seems too complicated to most beginners, because they think that they must memorize a lot of specifics. By understanding a few basic fundamentals, you can begin to utilize water temperature to locate gamefish.

Most of this information derives from three principles: (1) Much of the time, the water temperature is somewhere in the "comfort range," or temperature tolerance range, which is between 55 and 75 degrees for many species.[b] At times when the water temperature is near the middle of this comfort range, at the ideal

---

[b] The actual temperature tolerance range is different for each species of fish. The principles in this discussion are essentially the same for all species.

temperature of 60–65 degrees, for example, the location of fish will not be related to water temperature — it will be dictated by the currents, the natural bait (prey), and relative safety from predators.

(2) Often the water will be within the comfort range, but at the lower or upper limits, near 55 degrees or 75 degrees. At these times it is not the absolute temperature that determines the location of fish, it is the *relative* temperature. If the water is near the lower end of the comfort range, the fish will seek out water that is relatively warmer than the rest. They will be in the sunshine. If the water has been temporarily cooled by a blast of cold air, they will be in the deep holes in the bottom, which is farther from the cold air. If there is shallow water over black mud that absorbs heat, they will be in this shallow water that is warmer than nearby deeper water.

On the other hand, if the water temperature is near the upper limit of the comfort range, they will seek out the coolest water. They will be in the shade. They will be at the bottom of the middle of a pond in the summer, rather than the shallow banks where the sun heats up the water and the dark bottom. In saltwater, seatrout are found off the beach in summer when the inshore water is warm, because the deep ocean is a little cooler than the inland creeks, and the surf oxygenates the water. In winter these same seatrout are usually in the creeks, which are warmed by the sunshine on the black mud banks.

(3) During extremes of water temperature, outside of the entire comfort range, the water temperature will be the sole factor determining the location of fish. Cold-blooded animals have a wide tolerance for temperature, but they too can become hypothermic and overcome by the cold. On the other hand, when the water temperature is very high, the problem will be low oxygen. Recall that when you open a hot can of carbonated beverage, the gas fizzes out — a cold liquid retains its gas bubbles. Under these conditions, the fisherman should try to locate *oxygenated* water,

which is usually in cooler water, or where the surface is broken, or near the outflow of a turbulent side stream. Nothing else will matter to them if they are becoming hypothermic, or if they have insufficient oxygen. This is one of the few situations when food and safety from predators are not paramount. The other situation is when the instinct to breed overrides food and safety.

To summarize: When the water temperature is ideal (around 60 to 65 degrees), the fish will be wherever there is food and security, with no regard for temperature. When the water temperature is in the comfort range (55 to 75), but near the lower and upper limits (56 degrees, or 74 degrees, for instance) the fish will look for food and security in nearby waters with the best relative temperature. Finally, when the water temperature is too warm to contain adequate oxygen or too cold to support their metabolism (over 80 or under 50, for many species), the fish locate entirely on the basis of temperature, disregarding all other needs.

The influence of the weather and barometric pressure on fishing success has already been discussed in chapter 11, pages 193–195.

Successful fishing is all about probabilities and increasing your odds by balancing all of these factors in order to be at the right place at the right time. If, by no fault of your own, none of your ancestors were Native Americans or marine mammals, you must learn the basic biological needs of fish.

They don't wander around aimlessly like us. They don't goof off, or take vacations. They follow their instincts, all of the time, every day. Their instincts tell them to find food in a safe place, avoid extreme temperatures, and reproduce once a year.

In order for you to find them in saltwater, the tide table is a good place to start. The tide table helps you choose the best fishing days, around spring tides and strong currents. Even in your favorite fishing creeks, the predators aren't there all of the time, and the tide table is the key to timing your visit to likely feeding stations. Over shallow fishing grounds, the tide table and the Rule of Twelve can guide you to the perfect depth for the fish and for the fisherman. You want to find feeding fish, and the location of the natural bait and predator gamefish is largely determined by the tidal currents.

I cannot omit one last aspect of this topic: the solunar tables. Among both hunters and fishermen, you will find advocates of a system known as the solunar tables that claims to predict the *hours* of each day that are optimal for feeding behavior in both wild mammals and fish. I have looked into this enough to learn two facts: (1) There is absolutely no known biological or scientific basis for this theory. Even the earnest sportsmen who publish these tables have no explanation for why they should work. It is purely empirical and based on decades of observation. (2) Many highly respected hunters and anglers are totally convinced that there is some unexplained pattern of solar and lunar interaction that causes animals to feed at certain times of the day and night. I do not use the solunar tables, but I suspect that there *are* solar and lunar influences on biology of which science is totally unaware.

Another thing science cannot explain is that whenever two people are fishing together, using the same technique in the same place at the same time, one of them will usually be catching lots of trophy fish, and the other will be saying, "Looks like its your day, Ed."

Chapter Thirteen

The Constituents of the Tides on Earth. Synopsis of Tidal Influences

"Dwellers by the sea cannot fail but to be impressed by its ceaseless ebb and flow, and are apt...to trace a subtle relation, a secret harmony, between the tides and the lives of men."

Sir James George Frazier

One of my favorite professors in medical school was a crusty, down to earth old curmudgeon, whom we shall call Doctor B. He had been in medical practice since before the advent of antibiotics. He told us that in those days hospitals had special wards reserved for patients with osteomyelitis (bone infection). These unfortunate souls were sequestered together for a very practical reason. Their infections were attended by a terrible odor. The skeletal abscess would eventually find its way to the surface, creating foul smelling, draining tunnels called "sinus tracts."

One morning Doctor B.'s group of medical students and their professor found a new patient had been admitted to the "osteo" ward the previous night. When they pulled back the bed sheets to examine her wounds, they were amazed to see that the sinus tracts in her leg were clean and without any drainage or odor. The professor asked her how she had been dealing with her infection before she arrived at the hospital. The old lady announced that her mother had taught her to "stuff moldy bread in the holes." She said that this made them better, but she wanted the doctors to cure her completely. The young doctors were amused by this story, ignored it, and proceeded with the almost useless therapy they gave to all the other patients.

Several years later, Doctor B. picked up a medical journal and read that Edward Jenner had discovered penicillin, which is produced by the fungi that live in bread. When Jenner received the Nobel Prize, Doctor B. told us that the group of students who had been with "the bread mold lady" had no other choice but to "stand in a circle, and kick each other in the butt." True story.

Some of the tide generating forces are unexpected and improbable to say the least. I imagine there were many skeptics when Bernoulli first proposed that the tides were waves, 12,000 miles from crest to crest; and certainly when Laplace described the oceans as enormous agitated basins, with water from the previous tide sloshing back and forth to compound the following tide. And, the Coriolis force sending this motion off in a circular path, so as to make the sea level not level at all? And, what about that business of coastal waters that resonate like a tuning fork and accumulate energy like a swing set? And, the idea that what appears to be a simple flood and ebb against the shore is actually the sum total of dozens of individual constituent waves, which may reinforce each other like a rogue wave, or may interfere with each other and completely cancel out the tide for many hours at a time? As Lewis Carroll said in *Alice in Wonderland,* "Things just get curiouser and curiouser." The tidal wave, like most things in nature, is made up of many remarkable constituents; and they are integrated together so perfectly that we have always mistaken the flood tide for one simple wave pulled along by the moon — before we had this conversation, that is.

Now, before we finish our conversation, we should take a look at the complete list of constituents used by NOAA to compute the tide table distributed to the boating public. It is important to read this list, but it is *not* important to memorize it, or even retain any of the details. I suggest that you read it without much thought, disconnect your brain, and just absorb it like a frog absorbs things through his skin.

THE CONSTITUENTS OF THE NOAA TIDE TABLES

PRINCIPLE CONSTITUENTS:

M2 — The principle lunar semidiurnal constituent. M2 represents the rotation of the earth, as seen from the moon.

S2 — The principle solar semidiurnal constituent. S2 represents the rotation of the earth, as seen from the sun.

K1 — The lunisolar diurnal constituent. K1 represents the declination of the lunar and solar orbits to the earth's equator. When combined with constituent O1, they express the effect of the lunar declination. When combined with P1, they express the effect of the solar declination.

O1 — Together with K1, O1 expresses the effect of the lunar declination.

N2 — The larger lunar elliptic semidiurnal constituent. N2, together with L2 (the smaller lunar elliptic semidiurnal constituent), modulates the amplitude and frequency of M2, to account for the variation in the orbital speed of the moon, due to its elliptical orbit.

M4 — The major shallow water overtide of M2. M4 represents resonance of shallow water with the M2 tidal energy of the adjacent ocean.

SECONDARY CONSTITUENTS:

J1, M1, Q1 — The smaller lunar elliptic diurnal constituents. J1, M1 and Q1 all modulate the effect of the lunar declination to account for the moon's elliptical orbit.

K2 — The lunisolar semidiurnal constituent. K2 modulates both M2 and S2 to account for the declination of the moon and the sun, respectively.

L2 — The smaller lunar elliptic semidiurnal constituent. L2, together with N2, modulates M2 to account for the variation in the moon's orbital speed, due to its elliptical orbit.

M3 — The lunar tertiary, a shallow water constituent. M3 accounts for the deformity of the M2 tidal energy waveform that occurs in shallow water.

M6, M8 — Shallow water overtides of M2. M6 and M8 represent resonance of shallow water with the M2 tidal energy of the adjacent ocean.

2N2 — The lunar elliptic semidiurnal second-order constituent. 2N2 "fine-tunes" N2 (see above).

OO — The lunar diurnal second-order constituent. OO "fine-tunes" O1 (above).

P, P1 — The rate of change of the lunar perigee (P) and the solar perihelion (P1) with respect to the earth's longitudes.

2Q — The lunar elliptic diurnal second-order constituent. 2Q "fine-tunes" Q1.

R2 — The smaller solar elliptic constituent. R2, together with T2, modulates S2 to account for the variation in the earth's orbital speed, due to its elliptical orbit.

S1 — The solar diurnal constituent. S1 accounts for the earth's elliptic orbit around the sun.

S4, S6 — The shallow water overtides of S2. S4 and S6 represent resonance of shallow water with the S2 tidal energy of the adjacent ocean.

T2 — The larger solar elliptical constituent. T2, together with R2, modulates S2 to account for the variation in the earth's orbital speed, due to its elliptical orbit.

Lambda — The smaller lunar evectional constituent. Lambda, together with V2, u2, and S2, modulate M2 for the effect of the solar gravitation on the moon's orbital speed around the earth. As the moon orbits the earth, it may be traveling toward the sun or away from the sun's gravitation; this causes the moon to change speed as it orbits the earth.

Rho1 — The larger lunar evectional diurnal constituent. Rho1, like lambda, modulate M2 for the effect of solar gravitation on the moons orbital speed.

Mf — The lunar fortnightly constituent. Mf expresses the deformity of the M2 oceanic waveform due to shallow water hydraulics.

Mm — The lunar monthly constituent. Mm expresses the effect of irregularities in the moon's rate of change of distance and speed in orbit.

MSf — The lunar synodic fortnightly constituent. MSf accounts for the monthly variation in Mf, due to the earth's orbit around the sun.

Sa, Ssa — The solar annual constituents. Sa and Ssa account for the nonuniform changes in the sun's declination to the earth's equator.

MK, 2MK, MN, MS, 2SM — Account for additional interactions between lunar and solar orbits and gravitational fields.

Using only seven principle constituents, a tide table can be constructed that is accurate within a ten per cent margin of error. However, 37 constituents are represented in the tables distributed to the general boating public by NOAA, 114 constituents are included in their computer program to generate the more precise tables for any coastline affected by inland waterways and estuaries, and 396 constituents had been identified over thirty years ago by Dr. Authur Doodson.

Well into my research, I was still wondering where all of these natural forces were hiding among the orbits of the earth, moon, and sun. After all, there are only three astronomical bodies involved. I was aware that the earth orbits the lunar orbital axis, and also rotates on its polar axis, and the orbits of the earth and moon are elliptical, and both orbits change planes to the equator. All of this would doubtless require a few additional mathematical formulas to describe; but I couldn't imagine how this would require hundreds of constituent mathematical formulas.

Only after I prepared the above list of these constituents, did I fully understand the extreme complexity of this system. All of them are *nonlinear*. That is, they do not change the same amount with each passing hour — they are analogous to the acceleration of a vehicle which goes from 0 to 10 mph in the first 3 seconds, from 10 to 30 mph the next 3 seconds, and from 30 to 60 mph in the next 3 seconds. That complicates things.

In addition, they are all *simultaneous, and their effects are interdependent.* Since each body in the near solar system influences the others, every time one of them changes (and they are all changing, all of the time) this may simultaneously influence the tide caused by the other constituents. That *really* complicates things.

Let's first consider the celestial cycles from chapters two through four, the astronomical influences on the earth's tides. The oceanographers at NOS must program the computers with mathematical formulas that describe the planetary motion, the path of the celestial orbits, and the speed of the astronomical bodies in those orbits. I don't know about you, but I could not possibly describe an astronomical system with mathematical formulas. However, you and I *can* describe such a system with words.

(13-1) (a)   (b)

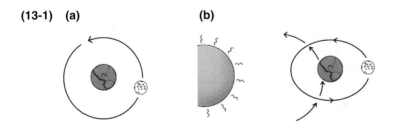

If I were describing illustration 13-1(a), I would tell the computer, "The moon travels around the earth at a constant speed in a perfect circle, which is always in the same plane. Lunar gravitation is uniform at the earth's surface."

I couldn't describe illustration 13-1(b), with any less than: "The moon travels around the earth in an elliptical path, while the earth travels around the sun in an elliptical path. Sometimes the moon is farther from earth than at other times, which diminishes the moon's gravitational field at the surface of the earth. Sometimes the moon is traveling away from the earth, which slows the moon's speed. This slows the rate of change of the lunar gravitation on the earth's surface. Sometimes the earth is closer to the sun, which increases the solar gravitational field on both the earth and the moon. However, the earth may be traveling *toward* the sun at the same time that the moon is traveling *away* from the sun in its orbit around the earth.

I am not too sure whether the moon is speeding up or slowing down when the earth is going toward the sun and the moon is circling back away from the sun? You're a computer. You figure it out. While you're at it, remember that during some seasons of the year the solar gravitation at the earth's surface is stronger or weaker, while the strength of the lunar gravitation changes every two weeks, and at some times they pull in the same direction, and at other times they pull in different directions, and therefore at some times you must add (or subtract) a larger (or smaller) solar tide to (or from) a larger (or smaller) lunar tide. And, also ... Oh *#*@#*, I give up.

(13-1) (c)

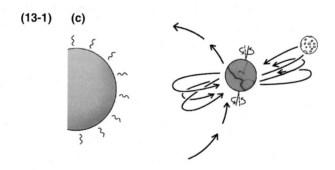

Now you try and describe illustration 13-1(c). Remember to tell the computer about the tilt of the polar axis that causes seasonal variation in the solar influence, and how the changing declination of the solar and lunar orbits to the equator may direct the crests of the semidiurnal waves over the same side of the equator on some days, or divert them over different hemispheres on other days. This may result in a diurnal pattern of solar influence on the tides in your hemisphere, while the lunar tidal waves are semidiurnal on the same day.

As if this were not enough, remember that we have not even considered the seiche effect over the continental shelf and within the basins of oscillation, which mixes the energy of successive tidal waves. In addition, we must add the hydraulic forces that elevate the tidal wave as it approaches the coast, first by the influence of the bottom topography and then by the coastal geography. Finally, remember that resonant co-oscillation and sustained forcing of tidal energy may further heighten the tide to extreme levels within a coastal basin, or the estuary effect may distort the tide cycle as the flood and ebb move inland. This is further complicated by the circular path of the tides around the amphidromes, and the displacement of the sea by coastal Kelvin waves. It should be no surprise that hundreds of mathematical formulas are required to describe this system.

The tide with all of its subtleties and nuances involve much more than the height of the flood, although it is the amazing extremes of the tide in places like the Bay of Fundy that capture our imagination. Oceanographers have used a number of different models to explain why the tides are higher at the coast than they are in the middle of the ocean, and why they reach extreme levels in some coastal basins.

We have explained the elevation of the tide over the continental shelf in a number of ways, including: (1) the reinforcement of multiple constituent waves arriving

simultaneously from the ocean, (2) the shallow water laws of hydraulics by which waves must slow down and gain height, (3) a standing wave over the continental shelf due to the mixing of incoming waves and reflected waves by the seiche effect, (4) the compression of tidal energy into a smaller volume as the tidal wave moves up over a gradually sloping bottom, (5) the compression of tidal energy into a smaller volume as the tidal wave moves into a receding coastline, (6) the oscillation of coastal waters that move up and down like a child's see-saw, and elevate the water at the head of a bay when the pivot point (node) is at the mouth of the bay, (7) the sustained forcing of more and more tidal energy into a coastal basin by successive tidal waves, when the period of natural oscillation in the bay is attuned to the period of the incoming oceanic tidal wave, akin to pushing a child on a swing set in rhythm with their established motion, (8) coastal Kelvin waves, which tilt the water higher on one side of a channel, (9) the increased height of the water at the outer periphery of the oceanic amphidromes.

This may be frustrating to some readers, who may ask, "Which model is it. Which one is the actual description of the tidal elevation at the coast?" The best answer is that all of them are valid and useful to understand the tides at the coast; but all of them are, in a sense, incomplete. You would have the same dilemma if you asked a group of scientists a question like, "What is a bird?" Because various experts approach scientific questions from different viewpoints, you would likely get several answers. An anatomist would describe how the bones are lightweight, and the animal is covered with feathers. A paleontologist might describe the evolution of an animal that can fly. A bird watcher would probably tell you about the variety of colors and vocalization by which it attracts a mate. An ornithologist might focus on the variety of species all over the earth. A behavioral biologist might emphasize their nesting habits and their migration. All of them are right, but the best answer comes from using all of this information.

Synopsis

The same thing can be said about our numerous explanations of the variety and complexity of tide patterns on earth. **We have alluded to: (1) the many distinct tidal waves circling the globe simultaneously, under the influence of the complex orbits in the near solar system, (2) the endless reinforcement and interference of these multiple waves, (3) the Coriolis force diverting the path of these waves, (4) the fact that different basins of oscillation selectively respond to different frequencies within the global tidal wave, (5) the fact that the water never comes back to a resting equilibrium before the next tidal wave arrives, (6) the seiche effect within coastal basins and over the continental shelf, which causes tidal intermixing, (7) the resonant overtides with shorter and shorter wavelengths that may be generated when the oceanic tidal wave moves into shallower and shallower water, (8) the deformity of the tide cycle due to the frictional delay of the ebb tide within estuaries, (9) the combination of independent tides (generated within a coastal basin) and co-oscillating tides (forced by the incoming oceanic tidal wave).** Again, all of these principles are correct; none of them are exclusive of the others; and all of them are essential to understanding the complexity of the tides.

Having learned so much about the tides on earth, we must be careful to avoid the possibility that we substitute our previous confusion due to ignorance with a new confusion due to all of this complexity. In this chapter, we have surveyed the 37 constituents used by NOAA to compute the tide tables. These are the cyclical (mostly astronomical) constituents that determine the frequency of the tidal waves. Then, we have reviewed 9 additional oceanographic, hydraulic, and geographic forces that determine the height of the tides over the continental shelf, and 9 earthbound forces that complicate the tide pattern at the coast. In order to weave all of these threads into whole cloth, all of this material is summarized in illustration 13-2, below.

(13-2)   The Harmonic Model of the Tides

Astronomy

Lunar
Elliptical orbit
Declination of orbit

Earth
Orbit of lunar orbital axis
Rotation on polar axis

Solar
Earth's elliptical orbit
Tilt of earth's axis to sun

↘ ↓ ↙

Celestial Harmonics
Reinforcement and interference of
annual, monthly, daily and hourly cycles

↓

Oceanography

12–18 inch ellipse (bulge) of oceans
Tractal forces
Coriolis effect — Amphidromes
Basin of oscillation — Seiche
Hydraulics over continental shelf

↓

24–36 inch tidal wave at the continental shelf

Hydraulics   ↓   **Geography**

Topography of nearshore bottom and the contour of the coastline
Shallow water wave hydraulics

↙ ↓ ↘

Amplification   Deformity of tide-cycle   Kelvin waves
by resonance   by estuary
sustained forcing

↘   ↙

0–50 foot tidal range at shoreline

↑

Meteorology

Barometric pressure   Wind effect   Temperature

Synopsis

As you scan down illustration 13-2, you see the chapters of *Beyond the Moon* in order from top to bottom. Throughout these chapters, we have repeatedly found examples of how various constituents of the tide interact with other constituents. In chapter three, we learned how the elliptical lunar orbit changes the importance of lunar gravitation and the lunar declination.

In chapter four, we noted that the earth's elliptical orbit of the sun influences both the solar gravitation at the earth's surface and the moon's orbit of the earth. Chapter five was entirely devoted to the constant reinforcement and interference of multiple simultaneous waves.

Chapter six described how the Coriolis force deflects the advancing tidal wave across the ocean. Then in subsequent chapters we found that this same deflection also causes damping of the reflected waves over the continental shelf (chapter seven), and affects the sustained forcing of tidal energy (chapter eight).

In chapter seven, we learned how seiche effect waves travel backwards to mix the energy of successive tides. This not only generates unique tides in some basins of oscillation like the Gulf of Mexico, but also elevates the height of the water over the continental shelf (chapter seven), and makes possible sustained forcing of energy into resonant bays (chapter eight).

In chapter eight, we found that at each point along the coast a unique combination of coastal contour and near shore topography interacts to create the height of the water. In addition, we learned that although co-oscillation depends on the width of the mouth of a basin and resonance depends on the length and depth of a basin, these parameters combine to determine the amount of resonant co-oscillation and sustained forcing that occur within each bay and harbor.

The deformity of the tide cycle that causes tidal pumping is the result of forces within an estuary (chapter nine). However, the resulting fortnightly tide cycle depends on the phases of the moon, far out in space (chapter four).

Over and over again we find endless examples of how **the celestial forces in chapters two through four combine with the ocean dynamics in chapters six and seven, which are amplified by the hydraulics and coastal geography in chapter eight, and which may then be modulated by an estuary in chapter nine, which in turn depends on the celestial forces, sending us back to chapters two through four again. And so there is a dynamic linkage, and a relentless interdependence between the astronomical, oceanographic, hydraulic, and geographic constituents of the tides on earth.**

The 17th century astronomer, Johannes Kepler, said, "To measure is to know." You can almost hear the 20th century oceanographers saying, "Look, Johannes, I am *trying* to measure it, but the damn thing *won't hold still*."

Now, let me tell you what is going to happen, after you close these pages. Someone is going to ask you why the tides are so unusual in their homeport, or at some place they recently visited. It is impossible to give them a short answer. Perhaps the best you can do is take them outdoors, and say, "See that cloud over there that looks like a wisp of cotton. Well, last week I saw one that looked like a huge anvil. For similar reasons, the tide in the Gulf of Mexico may not move at all for 8 hours, but the tide in the Amazon may race up the river in a ten foot wall of water. There are places on the coast of Europe where there are three high tides or four high tides each day. That seems no more unusual to me than this cloud over here, that is shaped like a camel, or that one over there, that looks like Abraham Lincoln." This is the best short answer, but they won't understand.

For many decades, the archaeologists who studied ancient Egypt struggled with the translation of hieroglyphs into modern language. Finally they discovered a stone tablet that listed many hieroglyphs together with associated pictures of objects, giving their meaning. This was called the rosetta stone, and it was the key to deciphering all ancient Egyptian writing. We are now going to use illustration 13-3 as such at tool — a sort of rosetta stone to decipher the tides.

We can reduce our probing of the tides to two questions: (1) Why are the tides different at different ports on the same day, and (2) Why are the tides different at the same place on different days.

Let's use illustration 13-3(a), below, to look at the constituents of the tides at two different ports on the same day. The annotations in red refer to the dominant tidal influences at a port called "Homeport." The blue annotations refer to the influences that dominate the tides at a second port, "Awayport." These are hypothetical locations — the Epilogue contains further examples of the practical application of *Beyond the Moon* at real coastal locations.

Both of these hypothetical ports are on the Atlantic coast of North America. Therefore they are in the same basin of oscillation. This means that the same tidal wave frequencies created by the same astronomical cycles will produce the same semidiurnal tide pattern in both ports. Likewise, the influence of all of the celestial cycles and the resulting celestial harmonics on this same date will be the same in both ports. After all, they are both in the same basin of oscillation and in the same solar system. Furthermore, both ports will realize the same elevation of the oceanic tidal wave out at the continental shelf (from about 18 inches, to about 36 inches).

(13-3) (a)   The Harmonic Model of the Tides

Homeport
Awayport

Astronomy

Lunar   *Earth*   *Solar*
Elliptical orbit   Orbit of lunar orbital axis   Earth's elliptical orbit
Declination of orbit   Rotation on polar axis   Tilt of earth's axis to sun

↘ ↓ ↙

Celestial Harmonics
Reinforcement and interference of
annual, monthly, daily and hourly cycles

↓

Oceanography
12–18 inch ellipse (bulge) of oceans

Tractal forces
Coriolis effect — Amphidromes
Basin of oscillation — Seiche
Hydraulics over continental shelf

↓

24–36 inch tidal wave at the continental shelf

Receding coast   **Hydraulics**   ↓   **Geography**   Straight coast
Steep bottom   Steep bottom

Topography of nearshore bottom and the contour of the coastline

6 ft. ←   Shallow water wave hydraulics

↙   ↓   ↘

Amplification   Deformity of tide-cycle   Kelvin waves
by resonance   by estuary
sustained forcing

9 ft. ←

↘ ↙

0–50 foot tidal range at shoreline

↑

Meteorology
Barometric pressure   Wind effect   Temperature   3 ft.

15 ft.

↓

Deformity of tide cycle

Continuing down the schematic, we arrive at the topography of the near shore bottom and the contour of the geographic coast. Notice that "Homeport" has a straight coastline and a steep near shore bottom. This will result in minimal change in the tidal range arriving from the continental shelf, and the high tide will be about three feet above the low tide at Homeport.

At Awayport, the coastline is receding inland and the bottom has a steep change in depth. This funnel shaped coast will elevate the tide higher than the three foot range at Homeport.

In addition, the length and depth of the coastal basin at Awayport produces a natural period of oscillation that is mathematically attuned to the semidiurnal frequency arriving from the Atlantic Ocean. This allows sustained forcing of tidal energy into this basin and further elevates the range of the tide. Let's say that the receding coast causes a six foot tidal range, and sustained forcing adds another nine feet, for a total range of fifteen feet at Awayport.

There is one other difference between these two locations, which are 500 miles apart. Near Homeport there are many rivers and shallow tidal backwaters, but there are none near Awayport. Therefore, at Homeport there may be a deformity of the cycle of flood and ebb tides. There will not be a tidal bore in any of these rivers because the tidal range is too small. However, there might be a stand in the tide, with no movement for many hours, or some delay in the ebb tide that is not found at Awayport.

Now let's turn to illustration 13-3(b) and again compare the different tide patterns occurring on the same date at two different ports — this time, a port on the east coast of North America, and a port on the Pacific west coast. We will call the Atlantic port "Homeport" (red) and the Pacific port "Remoteport" (green).

(13-3) (b) The Harmonic Model of the Tides

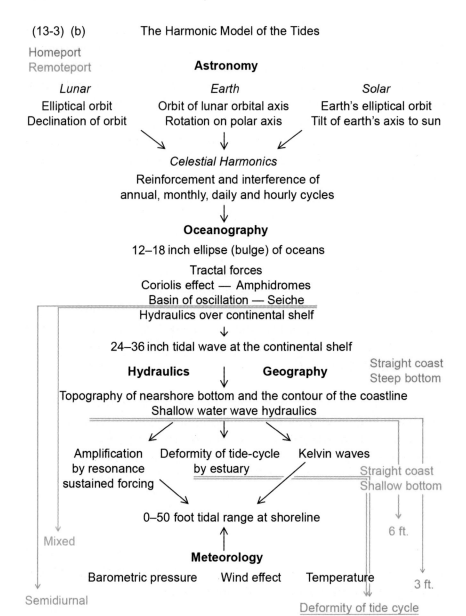

In this case, reading down the schematic, we find the first major difference is that they are in different basins of oscillation. Because of the different size and depth of the Atlantic and Pacific oceans, these basins will respond to different constituent frequencies of the tidal wave — it will be almost as if there were different celestial cycles generating the tides. Of course the celestial cycles will be the same, but the different basins will be attuned to different frequencies (see page 108). This will result in the greatest difference between the tides at these two ports: the tide pattern will be semidiurnal at Homeport, whereas the tide pattern will be mixed at Remoteport. This is a profound difference.

(13-3) (c)

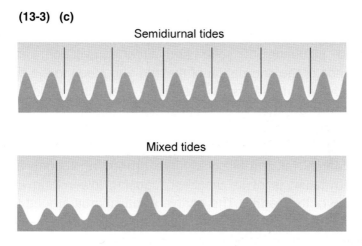

As you can see, **the selective response to different frequencies by different basins can orchestrate all of the other tidal constituents in a dramatic way.**

The coasts are straight at both locations, but the near shore bottom topography is shallower at Remoteport, and the tidal range will be higher there — let's say six feet, compared to three feet at

Homeport. Finally, the tides at both of these ports are influenced by the estuary effect, although this effect will be unique in each port, since no two estuaries are identical.

Now let's try using illustration 13-3 for analyzing variations in the tide at the same place *on different dates*. We will follow down illustration 13-3(d), below, on three dates: March 1st (green), August 1st (blue), and August 15th (red).

Since we are only dealing with one port, all of the parameters from oceanography down will be the same every day of the year. However, as the calendar progresses, the near solar system constantly realigns itself, and the distances between the celestial bodies constantly change. The concept of a lunar tide and a distinct solar tide becomes very important here, because of the changing influence of solar and lunar gravitation, and the significance of the declination of the sun and moon to the equator. Each of these parameters may reinforce each other at some times and interfere at other times. Let's "do the math" for these three dates at a hypothetical port, using illustration 13-3(d).

On March 1st, the elliptical lunar orbit is at apogee and the lunar gravitation is at its weakest, causing lower than average tides all over the earth. The earth's elliptical orbit around the sun is near aphelion, and the solar gravitation is also weak, reinforcing the lower than average tidal range on this date. The lunar orbit is at a minimal declination to the equator, and therefore the high tides are equal at our Homeport. The sun is declined slightly south of the equator. This diminishes the influence of the solar gravitation in the northern hemisphere, reinforcing the weak solar tide at Homeport. The sea level is also slightly lower than average in the northern hemisphere, because the solar radiation is not warming (and expanding) the ocean as much as it will in the summer. The summation of all of these constituents will be lower than average high tide levels and a small tidal range.

Synopsis

(13-3) (d) The Harmonic Model of the Tides

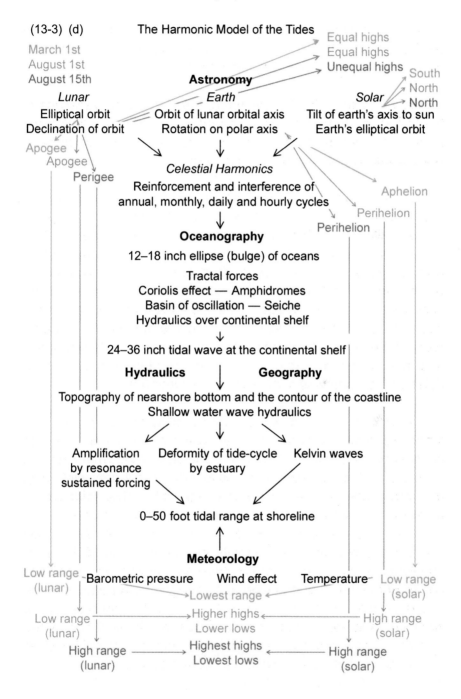

On August 1st, the lunar orbit is at apogee, and lower than average lunar tides would be expected. The earth's solar orbit is at perihelion, which has an opposite effect on the height of the water: higher than average solar tides. Now during spring tides, these large solar tides will be added to these small lunar tides, yielding moderate (average) tidal ranges. However, during neap tides, the large solar tides will be *subtracted* from small lunar tides, yielding very small tidal ranges.[a] On this date the lunar declination is minimal so that the high tides will be about equal each day. The sun is declined far from the equator over the northern hemisphere, and the thermal effect elevates the sea level throughout the tide cycle, during both spring tides and neap tides.

On this date, *during spring tides,* the overall effect of these constituents will be higher high tides and lower low tides than on March 1st, because stronger solar forces are now added to the lunar forces. Notice however, that during *neap tides* (quarter moons), the high tide will be lower (and the low tide will be higher) than on March 1st, because on August 1st, these stronger solar influences are *subtracted* from the weak lunar tides.

Continuing with illustration 13-3(d), on August 15th, the lunar orbit is now near perigee (15 days after apogee on August 1st), the lunar gravitational influence is maximal, and the lunar tides will be high. Of course the earth's solar orbit is still near perihelion (this cycle takes 365 days). On this date there is a coincidence of strong lunar and strong solar gravitation on the earth's oceans.

Furthermore, the sun is still declined over the northern hemisphere, the water is warmer and has expanded slightly, adding further to the height of the water.

---

[a] In illustration 13-3(d), the annotations such as "lowest range," "highest high," etc., refer to the spring tides. In order to determine the neap tides on these dates, it is necessary to apply the principles learned in earlier chapters, including pages 47, 48, 49, 52, and 150.

On this date, during spring tides, the tides are much higher and the tidal range is greater than on either March 1st or August 1st. In fact, these tides will be some of the highest tides of the entire year at Homeport, due to a confluence of multiple constituents that all reinforce higher tides. In contrast to this large increase in the spring tides, the neap tide range is essentially *the same* as on March 1st.[b] Again, this is because the lunar and solar tidal forces interact in a completely different way during full moon and quarter moon phases. This is a nice example of why the tides are not only *different* on different dates, **the tides are virtually unique every day of the year, because there are so many combinations of so many constituents**.

From now on readers should think in terms of a separate lunar tide and a solar tide when analyzing tide tables. In illustration 13-3(d), it was necessary to first consider whether the lunar tide was strong or weak on a given date, and then consider the strength of the solar tide. By consideration of these constituents separately, one can determine the result of either: (1) adding the solar tide to the lunar tide during spring tide phase, or (2) subtracting the solar tide from the lunar tide at neap tide.

Well, it seems that illustration 13-3 may indeed be the rosetta stone for deciphering tidal variations at different ports, and on different dates. Unfortunately, tattooing illustration 13-3 on your body is prohibited by copyright legislation.

---

[b] Larger solar tides subtracted from larger lunar tides may yield the same result as weaker solar tides subtracted from weaker lunar tides. For instance, if the lunar tide contribution was 11 units and the solar contribution was 5 units, the spring tide would total 16 units and the neap tide would be 6 units (11 − 5 = 6). If both solar and lunar tides increased to 15 lunar and 9 solar, the new spring tide would increase to 24 (15 + 9), but the new neap tide would still be 6 (15 − 9 = 6).

NOAA has settled on 37 constituents for the tide tables that are distributed to the general boating public. Why would they ever consider using 114 or even 396 constituents to construct a more precise table, since these other components each cause only a fraction of an inch of change in the water level? Recall our discussion of Chaos Theory in chapter ten. Each of these minor components has a minimal effect, but they are all occurring simultaneously in an infinite number of unique combinations.

Also recall our discussion of wave reinforcement and interference in chapter five. Most of the time the random combination of these small waves will cancel each other out by interference. Occasionally however, a unique combination will occur so that many of them will all crest at the same time. At those times, their mutual reinforcement could cause an unpredictable and significant anomaly in the water level. Thus, rare anomalies will occasionally occur, due to the reinforcement of many small constituents which temporarily become synchronous in the midst of their random combinations. This source of random error and unpredictability could be kept to a minimum by including all 396 of the constituents, with all of their cycles during 18.6 years (the longest period of any celestial cycle), measured at thousands of monitoring stations, located every mile along the coast.

As we have seen, there are other sources of difficulty in predicting the sea level and the tide cycles. Tectonic plates may shift the height of the continents. Climate change may affect the level of the oceans. The contour of the coast is eroded by currents, and the topography of the coastal seafloor is deepened by erosion and raised by sedimentation. The weather and freshwater runoff change the temperature, salinity, and viscosity of the seawater. NOAA does an admirable job of reacting to all of this and adjusting the tide tables; but **when it comes down to a race between an evolving planet and a governmental agency, nature is always ahead.**

You might be thinking, "Since computers are so fast and so available, why not include all of the cycles of all of the constituents in every tide table. Why not avoid any chance of anomalous tides that conflict with the tables?" First of all, we discussed in chapter ten that it is theoretically impossible to eliminate every trace of unpredictability from such a system — it simply cannot be done (page 175). Theoretically this *would* minimize the chance of unexpected tides, but as a practical matter it would be of no real benefit to mariners who use the tables out on the water.

Why not? The weather. Whereas random reinforcement of minor tide constituents may cause rare anomalous tides, and our evolving planet will occasionally change the physical parameters of tide constituents, the weather will often cause tides that conflict with the tables. The weather cannot be predicted by NOAA or anyone else when the tide tables are printed. **The effects of weather on the height of the water are much greater than the effects of random tide anomalies and much more common than changing physical parameters.**

Anyone who has read this far is now "the tide guru of the local marina." However, in spite of this newfound expertise, there will still be days on the water when the tides will vary slightly from the tide tables. My advice is: "Don't stake your life on it." Don't plan your boating itinerary too close to dangerous tides. On a practical level this means: add at least one extra hour of time, or one extra foot of water, when you are calculating your "float plan" near minimum acceptable depth for your boat, over known hazards. Remember that even the wizards at NOAA are at the mercy of the weather, and the complexity of nature.

Prudent mariners can rely on NOAA to select an appropriate number of constituents to provide tide tables that are reasonably accurate for their intended use. Nevertheless, an informed captain knows the tide may be at odds with the tables due to the complexity of nature, and a safe captain always keeps an eye to the weather.

Most mariners will confess that while venturing away from their homeport they have encountered tide patterns that did not make sense to them. The reason for this confusion is simple: on their ventures they carried with them a simplistic concept of the tides.

Now we understand the extreme complexity of this natural system. The astronomical forces that bind the solar system together, no less, provide the energy that sets the tides in motion. The rotation of the earth collides this tidal energy against the continents. Then, the dynamic forces of oceanography, the laws of hydraulics, and the infinitely irregular geography of the coasts act in concert (like so many musical instruments) to create the different tides found on long straight beaches and within receding coastlines, the unique tides in deep harbors and in shallow estuaries, the myriad patterns of the tides manifest in wide open bays and in seas with narrow straights, the odd tides in basins that oscillate, the huge tides in bays that resonate, the small tides on islands, and every tide swirling against the continents around the globe, all of them the logical result of natural forces now familiar to us.

All our lives we have accepted the fact that a cloud may have any shape. Feeling the shifting wind on our face, sensing the ever changing heat and cold, the rain and the mist and the dry air on our skin, hearing the leaves rustling in the breeze, seeing the smoke rise in chaotic turbulence, we never question the unique character of every cloud. Our intuition tells us that every cloud is a reasonable possibility.

I suspect that if we lived in the sea like the dolphin, and we could sense the many cycles of rhythmic tidal energy against our skin, then we would just naturally expect unique tides on every shore. Lacking that experience, we come to understand the complexity of the tides by learning about their many constituents. and how they interact.

Like an anatomist who dissects the whole into its many parts, we have taken a tide table apart and put it back together again. Like a scuba diver who has looked beneath the waves, we now see the ocean tide in a different way, with more depth and more detail.

What can we say about the need for empirical data gathering to predict the tides, our inability to make a tide table by pure calculation? According to the German philosopher, Goethe, "Thinking is more interesting than knowing, but less interesting than looking." If Goethe is right, then it's better that some things are just beyond our computer models. I'm glad that NOAA can provide us with adequate tide tables for safe boating. At the same time, it's good to know there is still a little mystery left in the world.

Complexity does not make life more difficult; complexity makes life more interesting. Complexity functions in nature to create stability. Complexity is the source of beauty in nature. An anonymous native American once said, "All my life I traveled through skies of great mystery, on winds of high adventure." Native Americans understood nature. Those of us who travel on concrete and view nature through a windshield, can study the tidal constituents in order to interpret the tide tables, but we cannot really master the primal force that is the tides. I wouldn't go to sea with anyone who thinks they can.

Now when we venture to faraway harbors, we will not be surprised to find unique tide patterns. Tide tables that were the most confusing will now, instead, seem the most interesting. Of course, there will still be times when we encounter tide patterns that defy analysis by the amateur oceanographer. Occasionally, the waters will be so turbulent and opaque that we cannot see which forces of nature are responsible. Even then, **we will understand that every tide pattern is a reasonable possibility, just as we know that a cloud may have any shape.**

Chapter Fourteen

Epilogue

"In the world of the blind, a one-eyed man is king."

Desiderius Erasmus

Early on, it was my intention to make it possible for sailors, yachtsmen, weekend boaters, saltwater fishermen, and mariners of all sorts to analyze any tide table on any coast, and reckon for themselves the explanation for any tide pattern they might encounter. As my research progressed, it became obvious that this was tantamount to promising a class of biology students that if they studied hard they would be able to explain how an amoeba works. Of course, they would know a lot about amoebas and one-celled animals in general, but probably not enough to make one at home.

It has been said that the secret of happiness in life is to have low expectations. If you were hoping to know for absolutely certain exactly why every nuance and detail of each tide table was so, then you may be disappointed. On the other hand, if you have been frustrated by the variety of tides at different locations and confused by the changing tide patterns on different days at the same place — if you would be satisfied just to feel comfortable with the tide tables wherever you traveled, and knowledgeable about the tides in general — then welcome aboard. You have realistic expectations and you are going to be satisfied.

Let's be philosophical about it. None of us understands everything about any subject of scientific inquiry. It should be enough to acquire a thorough working knowledge of the tides on earth, a subject of extreme scientific complexity and great practical importance.

To demonstrate that our newfound working knowledge really works, I am going to use a personal example of how the information in this book enabled me to analyze real tides and tide tables. I am fortunate enough to live on the northeast coast of Florida and spend my summers on the coast of Maine. Once a year I take a fishing trip to the outer banks of North Carolina. Although they are all on the same east coast of the same continent, the tides in these three places are quite different in several respects. This used to be a mystery to me, but now these differences are reconciled. Let's look at how this can be done.

You can make an educated guess about tidal variations with nothing more than a tide table that includes the lunar phases, a chart of the coastline, and a copy of illustration 13-2. However, the minimum information required to make an in-depth analysis of a local tide pattern includes: (1) the tide tables for one year, (2) a calendar of the lunar phases, (3) information on lunar and solar declination, and lunar and solar distance from the earth, (4) a map of the coastline, including several hundred miles north and south of your port, (5) a bathymetric chart of the local waters, showing offshore depth contours, (6) an oceanographic chart of the amphidromes that affect your coast, (7) local knowledge of the inshore tidal zone (rivers, estuaries, etc.), and (8) access to a library or the internet so that you can learn if oceanographers have identified a coastal basin with resonant co-oscillation, or coastal Kelvin wave influence, or a unique seiche effect near your region of interest. (9) Sometimes it is useful to look at tide tables for other ports for comparison. This can often confirm or disprove your explanations, and may or may not be necessary.

Solar declination is evident from the seasons of the year: in the northern hemisphere, the sun is declined north of the equator in the summer, etc. Lunar declination is very complex, changing in two different major cycles with periods of one anomalistic month, and 18.6 years, respectively. However the lunar declination can be deduced from looking at the 12 month tide tables for a port that has: (1) very simple semidiurnal tides, and (2) a sizable tidal range (northeast Florida is ideal). By finding the weeks with the most unequal daily high tides, you can assume lunar declination is the greatest. Weeks with very equal daily high tides will correlate with the least lunar declination. Then, you can apply this information about lunar declination to your analysis of other ports, where the tide table is more complex (the timing of the declination effect is the same everywhere on earth).

Solar perihelion is on January 2nd, and solar aphelion is on July 2nd. However, you will find that as a practical matter **solar tidal influence is related to the solar declination (and thermal expansion) more than the solar distance.** At my homeport, the highest tides of the year are near to the summer (solar declination over the northern hemisphere), rather than the winter (solar perihelion).

Lunar perigee and apogee occur in monthly cycles. Remember that they have nothing to do with the phases of the moon (full moon, quarter moon, new moon). You will have to dig these dates of lunar perigee and apogee out of the astronomy literature, for a precise analysis. They cannot be deduced from simple semidiurnal tide patterns, because **the influence of the lunar phases is greater than the influence of lunar distance.**

The first thing I look for is the basic tide pattern: semidiurnal, diurnal, or mixed. A semidiurnal pattern tells me that I should think mostly about lunar influences. Mixed patterns have a greater solar diurnal influence (page 108). Mixed patterns are also associated with the changing lunar declination (page 67). Pure diurnal patterns are uncommon and probably indicate the basin has a major seiche effect, which is canceling the semidiurnal tidal wave (page 104).

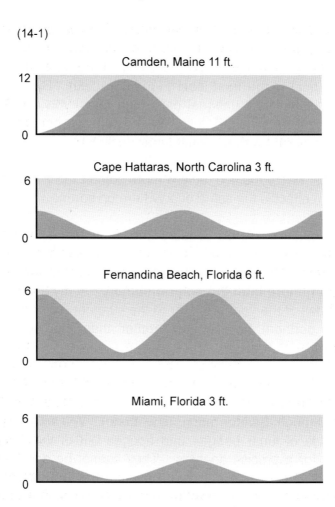

(14-1)

Camden, Maine 11 ft.

Cape Hattaras, North Carolina 3 ft.

Fernandina Beach, Florida 6 ft.

Miami, Florida 3 ft.

The tides are semidiurnal at all of my ports of interest. See illustration 14-1. This greatly simplifies things since they are all under the influence of the same basin of oscillation, so that the same astronomical cycles dominate all of these coastal locations.

As an additional comparison port I have chosen Seattle, Washington, because it is far away on the opposite coast, and because I intend to go there soon to introduce myself to the pacific salmon heading up the Columbia River. The Seattle coast has a mixed tide pattern. This will complicate a comparison between the Florida and Seattle tides, because many of the differences would be related to a different basin of oscillation (the Pacific Ocean). But to make this interesting, after we finish the east coast comparisons, we'll take a hypothetical trip to Seattle, and we will do an analysis of these east and west coast ports.

In the meantime, our comparison of three ports on the east coast has already provided an important clue — all of the tide tables reveal semidiurnal tides. **Most of the time, comparisons involve tides on the same coast, within the same basin of oscillation, so that the astronomical influences are the same. This means that all of the differences will be due to earthbound influences: oceanography, shallow water hydraulics, and coastal geography** (page 57).

The second thing that I look for is the average tidal range. Which port has the highest high tides, and how great is the difference? Notice that the tides range about 3 feet at the outer banks of North Carolina, 6 feet in northeast Florida, and 10 feet on the midcoast of Maine. Having spent my adult life in diagnostic medicine, I am dedicated to the principal that "common things are common." In other words, when you hear hoof beats, you don't think of zebras. Therefore, **I always assume that the tidal range difference is probably related to the contour of the coastline (page 122). Then if there is some discrepancy, I look for the explanation in the near shore topography (page 125). Next, if there is a very large tidal range involved (over 10 feet), I consider the other causes of extreme tides: resonant co-oscillation with sustained forcing of tidal energy into a coastal basin (page 136), or coastal Kelvin waves (page 84). If there are no extreme tidal ranges, I can eliminate these forces from the tide analysis.**

The major reason that the tides are twice as high in northeast Florida as they are at the outer banks of North Carolina is that this coastline recedes inland toward the west, in a slightly funnel shape (review page 124). Northeast Florida is at the point where the most tidal energy entering this entire coastline from Cape Hattaras, North Carolina to Miami, Florida is compressed into the center of this funnel shaped coast. In fact, there is an almost perfect correlation between the tidal range and the location of each point along this stretch of the coast.

Although we have our explanation for the tidal range differences from the outer banks down to Miami, let's look at the near shore topography anyway. I have done this, and this exercise illustrated two important principals of tide table comparison. First, on the larger scale of this entire 800 mile long coastline, it turns out that the near shore bottom topography is quite steep near the shoreline at both Cape Hattaras and Miami, whereas the near shore depths are relatively shallow in northeast Florida. This would tend to cause higher tides in northeast Florida compared to Cape Hattaras and Miami (page 125). In this case both the coastline and the bottom topography cause the same tide patterns. Which one has the greatest influence? It is possible to answer this question by gathering detailed data and employing advanced mathematics. However, I don't really need to know this detail. This illustrates that much of the uncertainty about tidal influences is of little practical importance. If there are two explanations for the observed tide pattern, it is not really necessary to know whether one contributes 60%, and the other 40%, or vice versa.

The second observation comes from a smaller scale analysis of short segments along this coastline. Although there is a progressive increase in the tidal range as you go south from North Carolina to northeast Florida, you can find inconsistencies such as the higher range along the Georgia coast, compared to the northeast Florida coast. This is unexpected since northeast

Florida is located closer to the *middle* of the receding coastline between Hattaras and Miami and should have a higher tidal range on the basis of the coastal contour. The most likely explanation is a difference in the topography of the near shore bottom.

In fact, the offshore depths are very shallow for many miles off the Georgia coast, compressing the volume of the advancing tidal wave and raising the water level there — bathymetric charts show the ten-fathom depth contour line is about 60 miles offshore of Savannah, and about 45 miles off Brunswick, Georgia. By contrast, the ten-fathom depth contour line is about 30 miles offshore at Jacksonville, Fla., and only 20 miles off Daytona Beach. This greater depth near north Florida ports absorbs some of the tidal wave volume and explains the larger tidal range at the relatively shallow Georgia ports. By noting both the coastal contour and the near shore topography, we can explain all of the tidal ranges along this coastline. However, before we head up to Maine, we should confirm our conclusions by running through a checklist of other possibilities.

There are other possible influences on the ports that we just compared: (1) Is there a seiche effect at one port that doesn't affect another? This is not likely when they are all in the same basin of oscillation (page 108). (2) Does the distance to the nearest amphidrome (in the Caribbean sea, in this case) have a significant effect? If two ports are on different co-tidal lines or co-range lines, this will contribute to the time of arrival and the height of the tide (page 81). These effects are often lost amidst the other larger forces that occur over the continental shelf, but they may explain a difference between two ports in cases where all of the other constituents are equal. (3) Does tidal pumping within a shallow estuary distort the tide cycle at one port and not the other (page 150)? In our case, there are estuaries and similar backwaters near all of the ports from North Carolina to south Florida, and this effect is fairly uniform throughout this coastline.

(4) Is there resonant co-oscillation and sustained forcing affecting the range of the tide (page 136)? This is not usually a consideration unless extreme tides are involved. (5) Is there any evidence of coastal Kelvin waves (page 84)? This is expected within confined channels along the coastline — checking the local maps and charts I see no reason to suspect that this might explain any local elevation of the tidal ranges along this coast.

The above paragraph can be replaced by referring to illustration 13-2, the one page synopsis of *Beyond the Moon*. Mariners who are interested in comparing the tide patterns in different ports can use this synopsis as a checklist of tidal influences.

In the summer I travel from Florida to the mid-coast of Maine. The tides there range about 10 to 12 feet at many ports. Camden, Maine, is on the coast of Penobscot Bay, a large basin which is part of the Gulf of Maine. The greatest tidal range on earth is in the Bay of Fundy, which is the farthest inland extension of this same Gulf of Maine. The reason for the rather extreme tidal range at Camden is sustained forcing of tidal energy into the Gulf by successive tides: the same forcing of the natural oscillation that occurs within the Bay of Fundy. Within the Gulf of Maine, the closer each bay is to the Bay of Fundy, the more it is influenced by this resonant co-oscillation and sustained forcing, and the greater the height of the water at high tide. See illustration 14-2. This is the major difference between these tides and the tides in Florida, where there is no significant resonance.

It is interesting that the tides on the coast of Maine would have a *lower* range purely on the basis of the geography and topography — in Maine the bottom drops off very quickly near the shore, which would make the tidal range less than in Florida, which has a relatively shallow near shore bottom. When you find that this is not the case, you must look for some other explanation. In this case, resonant co-oscillation and sustained forcing dominate the tide pattern on the coast of the Gulf of Maine, and coastal geography and topography play a less important role. Notice that sometimes two tidal influences may push the height of the water in the same direction — coastal contour and bottom topography along the coast from Cape Hattaras to Miami, both cause higher tides in northeast Florida. At other times, two tide constituents push the tide pattern in opposite directions. Bottom topography in Maine makes the range lower; resonance in Maine makes the range higher.

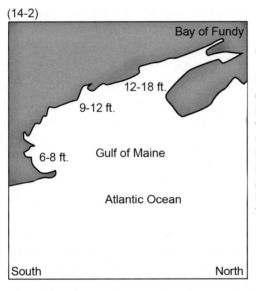

(14-2)

Gulf of Maine tides:

Lowest tidal range is ocean. Highest range is Bay of Fundy. The tidal range within the Gulf of Maine is influenced by the resonant co-oscillation and sustained forcing in the Bay of Fundy. The Gulf tidal range is 6–8 ft. in the south, 9–12 ft. in the middle, 12–18 ft. in the north.

If you look for subtle differences in the tide patterns of these two states, you can see a greater influence of tidal pumping within the salt marsh estuaries along the southeastern coast of the United States, compared to the rocky coast of Maine. For instance, if you look up the time of high and low tide *at the coast* in Florida, and then you want to know when these same tides will arrive at a place that is three miles *upriver*, you may find that the high tide arrives 30 minutes later, but the low tide arrives 60 minutes later upriver, in Florida. In Maine, both the high and low tides will have almost the same delay compared to the coast. This subtle tide feature depends on the amount of deformity of the tide cycle due to tidal trapping in the shallow vegetated backwaters.

If you look at similarities between the tide patterns in these two states, that are over a thousand miles apart, you find that tidal influences such as the declination of the lunar orbit (page 33) are the same. On January 14th, 2004, the two high tides in Camden, Maine, are only about two inches different. The two high tides in Fernandina Beach, Florida, are only about one inch different on this date. On January 21st, 2004, the two high tides in Camden are about 18 inches different because of the lunar declination, and the two high tides at Fernandina Beach are about 12 inches different on this date. If you make a habit of making such comparisons between your two ports of interest, it will be very useful to sort out which influences are the same at both locations, and allow you to delete them from the list of possible explanations for observed tide table differences.

We could go on and on listing differences and similarities between any two ports. In every case we find that the **differences on the same date at different ports are related to local features**: geography, near shore bottom contour, the local basins of oscillation (same or different at the two ports), the presence of a seiche, or coastal Kelvin waves, or major resonance within a coastal basin, the distance to the nearest amphidrome, and the relative importance of tidal trapping inshore. Why the tides are different on different dates at the same port is another question altogether. Remember that **differences in the same port on different dates are due to astronomical constituents**.

Now let's travel to Seattle, Washington, to keep our date with the pacific salmon heading for the Columbia River. The tide table is much less orderly here, and has bizarre variation compared to the Atlantic coast. See illustration 14-3. In Seattle, on January 14th, 2004, the tidal range in the morning is 9 feet, whereas the tidal range in the afternoon is only 5 feet. On January 21st, the tidal range is only 5 feet in the morning, and the afternoon tidal range is 15 feet. To restate this same information, on the 14th, the difference between the two high tides is 9 feet, and the difference between the two low tides is 5 feet. On the 21st, the difference between the two high tides is 5 feet, and the difference between the two low tides is 15 feet.

On the Atlantic coast we are accustomed to a difference in the two high tides on the same day, when there is a large declination of the lunar orbit to the equator. However, on the Atlantic coast the difference in high and low tides is synchronized and more consistent from day to day. Here on the Pacific coast, there is not the same linkage of the high and low tides — the pattern in Seattle does not appear to be related to the lunar declination. What is causing this pattern, so unfamiliar to Atlantic coast residents? The answer is that we are now in a different basin of oscillation, the Pacific Ocean. This Pacific basin is much larger and deeper than the Atlantic basin, and it is attuned to the longer wavelengths of the solar diurnal tides (page 108). At the same time, the lunar semidiurnal tides arise from the stronger lunar gravitation. Therefore, the Pacific tide patterns are "mixed tides," the complex result of constant reinforcement and interference between solar diurnal and lunar semidiurnal cycles.

The Atlantic tide patterns are a Strauss waltz; the Pacific tide patterns are Scott Joplin syncopated jazz. It is the influence of these different basins of oscillation that accounts for the major differences in the tide patterns found on the east and west coasts of North America.

(14-3)

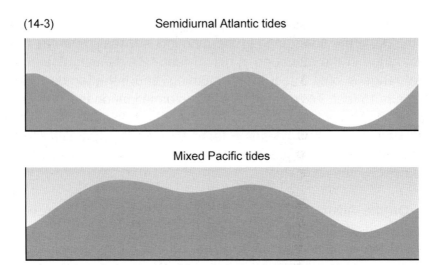

The Seattle tides are also influenced by the geography and topography of the coast, and like any other port, they are subject to the possibility of sustained forcing in a bay, and tidal pumping inshore. However, these influences are very hard to see beneath the irregular tide patterns that result from all of that solar and lunar reinforcement and interference. Suffice it to say, **analysis of the tides is often obscure where there is a mixed tide pattern.**

This will usually be the case whenever you travel to a new coast, or wherever you sail on your next voyage. **All of the astronomical, oceanographic, hydraulic, and geographic influences are included in every tide pattern, but they are often hidden beneath the larger waves of some locally dominant influence.** As we have seen, it is hard see the influence of the near shore topography in Maine when it is overwhelmed by the sustained forcing of natural oscillation that creates the Bay of Fundy tides. Likewise, on the Pacific coast, it is hard to hear the quiet conversation between the many tide constituents in a room filled with syncopated jazz.

This does not imply that the more subtle tide influences are not important. I was once asked why the high tide at a certain port arrives at different hours of the day as you proceed up and down the coastline. All of the major forces were about equal at the different ports in this place, and the answer was that the tidal wave travels counterclockwise in the northern hemisphere due to the earth's rotation, and this may sweep the high tide up or down a coastline as the tide progresses hour by hour. **A major tide influence may overwhelm all of the others, but they are still there. When all of the major forces are equal, a more subtle influence may explain the tide pattern.**

(14-4)

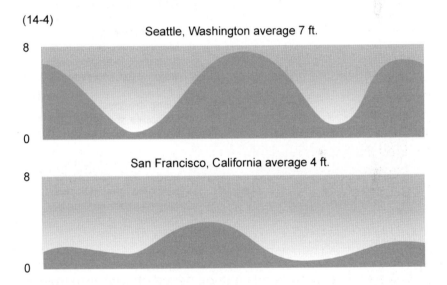

In spite of the relatively erratic patterns of the mixed tides in the pacific basin, we can still make comparisons between Seattle and other ports along the west coast. The average tidal range in San Francisco Bay is half as large (4 feet) as the average range in Seattle (7.6 feet). See illustration 14-4. The average range in Cooks Inlet, Alaska is almost twice as large (13 feet) as Seattle.

Bathymetric (depth) charts of the Pacific Ocean show a progressive shallowing of the water over the continental shelf as you proceed from Mexico to Canada. See illustration 14-5. There is a uniform correlation between the depth along the west coast (shallower as you go from San Francisco to Seattle), and the average range of the tide (higher as you go from San Francisco to Seattle). Indeed, this same trend includes the entire continental shelf from Central America to Alaska — as the near shore depth is less, the tidal range is greater.

(14-5)

Bathymetric (depth) chart:

Just off the coast of Mexico, there is a strip of dark blue, indicating deep water at the shoreline.

Along the coast of California, the lighter blue area is more shallow.

At the U.S.–Canadian border, there is a circular pale blue area, indicating very shallow near shore topography.

In Cooks Inlet, Alaska, the natural period of oscillation from the mouth to the head of the basin (and back to the mouth) is 12 hours, 15 minutes. This is almost exactly the same as the period of the principal lunar semidiurnal constituent (period of 12 hours, 26 minutes). This creates resonance between the oceanic semidiurnal tidal wave and the natural oscillation of the water in this basin, and promotes sustained forcing of tidal energy into Cooks Inlet (page 136). This is why there are tides of over thirty feet at the head of this bay, at Turnagain Arms. The resonant co-oscillation in Cooks Inlet, and the absence of resonance at Seattle, is clearly the explanation for the difference at these two ports. Even the obscurity of the mixed tides on this coast cannot hide this fact from us.

Finally, should you encounter a tide pattern that is totally obscure, along any coast or within any basin, do not think that you have missed something in your education on the tides. There is no other major tidal influence that has been withheld from you. Unique tide patterns often arise simply because there are so many constituents involved, and occasionally they will combine in a unique way, so that several of them will reinforce each other to produce an unexpected pattern, or a surprising high (or low) water level.

This is a bona fide explanation for many unusual tides, but it reminds me of how doctors declare that, "it's a virus," causing your fever (headache, cough, skin rash, etc.). Reinforcement is the "virus" of tide analysis — hard to prove, but often the correct diagnosis, after you exclude all other possibilities. **Reinforcement of a unique combination of influences should always be considered to explain otherwise inexplicable tides. Interference of two simultaneous influences may delete or obscure some influences, and lead to difficult interpretation of the tide pattern.**

The synopsis of *Beyond the Moon,* found in illustration 13-2, is a handy reference guide to help you analyze real tide tables and real tides found in real ports. With few exceptions you can be confident of the dominant tidal influence wherever you travel. Major differences in tides at different ports can almost always be explained by the oceanographic, hydraulic, and geographic tidal influences listed in this guide. The explanation for changing tides in the same port will be found somewhere in the astronomical forces and celestial cycles. The only other thing you need to know is this: When answering questions from the local residents about their tides, always exude confidence, use big words (amphidrome, lunar declination, resonant co-oscillation), and never forget that, "In the world of the blind, a one-eyed man is king."

Tide tables are available on-line at no expense, at the address of government agencies such as NOAA. Many nongovernmental web-sites also provide both tide tables on the internet and sell software for the serious mariner or fisherman to use at home. Using these resources, illustration 13-2, the synopsis of *Beyond the Moon*, and the other suggested charts and local information, **you should be able to at least make some sense of the myriad tides around the world. In order to do this you must look up into space, look beneath the waves, look along the shoreline, and down into the tidal backwaters. You must consider the orbits of the near solar system, the rotation of our earth, the dynamics of her oceans, the hydraulics over the continental shelf, and the geography of the continents.**

Usually this will be simple. Occasionally it will be nearly impossible. But even then, you will know that the constituents of large tides are the same as the constituents of small ones, and the constituents of obscure patterns are the same as the constituents of simple ones. This is not only true of the tides, it is seen throughout nature. The constituent known as carbon may be organized into a lump of coal in one place, or a diamond in another place. In one configuration, the constituent known as silicon will form a grain of sand. In another configuration it may compose a pane of glass, or even a computer chip. **The secret behind the infinite variety of the tides is found within the finite list of constituents common to them all.**

Practical Definitions

In keeping with the format of this book, these definitions are the common sense, practical usage of these terms. More scientific information in *Beyond the Moon* is referenced by page numbers, in the Subject Index. Expanded definitions are available on the internet at: www.co-ops.nos.noaa.gov., tide and current glossary.

Amphidrome: A circular pattern of oscillation in the oceans, counterclockwise in the northern hemisphere, and clockwise in the southern hemisphere, due to the deflection of the tidal waves because they are moving over the surface of a rotating sphere.

Amplify: To increase the magnitude (see amplitude below).

Anomalistic month: The time required for the moon to orbit the earth 360 degrees, and return to a position exactly over the same longitude on earth — 27.5 days. The difference between an anomalistic month and a sidereal month (one 360 degree lunar orbit) is due to the rotation of the earth during the lunar orbit.

Anomaly: A measurement which does not conform to theoretical expectation (the model). Anomalies arise because of either : (1) an unsuspected change in the parameters of the model, or (2) an unpredictable combination of events.

Aphasic: When two simultaneous forces are not synchronized — for example, when one of them is at a maximal strength, and the other is at a minimal strength — they will interfere with each other.

Amplitude: Scientific term for the greatness of size or strength, the magnitude.

Aphelion: The annual point in the earth's elliptical orbit around the sun, when the earth is farthest from the sun.

Apogee: The monthly point in the elliptical lunar orbit, when the moon is farthest from the earth.

Barometric pressure: The weight of the atmosphere from outer space down to sea level — normal pressure at sea level is equal to the weight of a 29.92 inch column of mercury.

Basin of oscillation: In the terminology of oceanographers, each geographic basin that determines the frequency and period of oscillation of the water in that region of the earth. This may be a coastal bay, a gulf, sea, ocean, or a portion of an ocean on one side of a submarine mountain ridge.

Calendar month: One twelfth of a year — 30.25 days — not related to the tides.

Celerity: The speed that a wave moves over the bottom.

Centrifugal force: The apparent force (outward) away from the axis of motion, exerted on an object moving in a circular or curved path, due to the momentum of the object. In the case of an object which is either stationary or moving on the rotating earth, physicists must formulate an apparent centrifugal force in order to explain their observations using Newton's Laws of Motion.

Chaos theory: A body of scientific inquiry that emphasizes the effect on dynamic systems due to the complex interaction of multiple, simultaneous, interacting constituents. One of the major conclusions derived from chaos theory is that whereas these individual constituents are governed by relatively simple laws of physics and mathematics, the overall outcome of their interaction is unpredictable, because the constituents can interact in so many different combinations. Therefore, nature is, at the same time, both deterministic and unpredictable.

Constituents of the tide: The harmonic model of the tides recognizes the simultaneous contribution of many different forces. Those component forces (constituents) that are cyclical, and can be expressed mathematically as a sinusoidal (~ shaped) curve on a graph, are included as constituents of the tidal wave that we observe at the coast and measure at tide monitoring stations.

Continental shelf: The seafloor between the shoreline and the abrupt vertical drop into the abyss of the ocean.

Cooscillation: In the terminology of oceanographers, the condition in which two distinct bodies of water exchange energy by a relationship between their natural periods of oscillation. This may be two geographically separate basins, or two adjacent zones of the same basin, with very different depth.

Coriolis force: The apparent (sideways) force exerted on objects moving over the surface of a rotating sphere, such as the earth. This apparent force is the cause of the circular motion of weather patterns in the atmosphere, as well as circular ocean currents and deflection of the tidal wave. It is towards the right in the northern hemisphere, and toward the left in the southern hemisphere.

Corange lines: Lines connecting points of equal tidal range in a region of the ocean.

Cotidal lines: Lines connecting points where the high tide arrives simultaneously in a region of the ocean.

Declination: The **lunar declination** indicates the angle between the lunar orbit around the earth, and the equator of the earth. The **solar declination** indicates the angle between the earth's orbit around the sun, and the earth's equator.

Diurnal tide patterns: Tide patterns which only have one recognizable high tide each 24 hours.

Dynamic theory of the tides: Replacing the Equilibrium Theory of the tides, this model recognized the influence of the rotation of the earth, the dynamics of the oceans, the hydraulics over the continental shelf, and the geography of the continents, as they affected the tides observed at the coast.

Ellipse: An oval shaped, flattened circle. A circle has one central axis and one radius. An ellipse has two axis points called foci. When you add the lengths of the two radii extending from the two foci to any point on the circumference of the ellipse, the sum of the two radii is always the same.

Empirical: The empirical method relies on observation and derives its conclusions from measurement, without regard for theory or mathematical calculation.

Equilibrium: In the terminology of physicists, equilibrium is a condition in which all previous added energy has dissipated, and a stable, resting state has been reestablished.

Equilibrium theory of the tides: Historically, the first unified scientific theory of the tides on earth, which assumed that the tides were solely based on the gravitational attraction of the moon on the earth's oceans.

Establishment of the tide: The time of delay between the passage of the moon directly overhead and the arrival of that lunar tide. Also called the **lag of the tide,** or **age of the tide**. There is also establishment of the solar tide.

Estuary: A shallow natural basin of water and wetlands located in the tidal zone, near the coast. Estuaries constantly exchange water and sediments between the land and sea, and are the "nursery" of the food chain, providing vegetation for the microscopic creatures, hatchling fish and crustations, and the coastal fishery.

Fathom: Originally, the length of rope between the two outstretched hands of a sailor lifting a piece of lead up from the bottom, in order to measure the depth. A fathom is now six feet.

Frequency: The number of complete cycles during a period of time.

Geostrophic forces: The influence of earthbound forces on the tidal wave: the rotation of the earth, the movement of the tidal wave over the shallower continental shelf, the interruption of the tidal wave by the continents, etc.

Gravitation and **Gravity:** Gravitation is the force of attraction between two objects with mass — the force that Newton described. Gravity is a more specific term for the force(s) that determine the weight of an object on earth. The weight of an object in the earth's sphere of influence is determined by the gravitation of the earth, the gravitation of the object, and the centrifugal force related to the earth's rotation.

Half wave oscillator: A basin of water, such as a lake, with a length that is exactly one-half the wavelength of its natural oscillation, placing the node of the wave motion exactly in the center of the basin.

Harmonic model of the tides: A model of the tides on earth, based on the fact that the observed tidal wave traveling around the globe (and the observed cycle of flood and ebb at the coast) is actually composed of a finite number of independent cyclical constituents, arising from the periodic movements in the near solar system as well as cyclical phenomena in the oceans and continental shelfwaters.

Harmonic analysis: A mathematical method by which the individual contribution of many constituent tidal waves are extracted from the measured height of the water, the observed tidal wave.

Harmonic constant: A mathematical expression of one of the cyclical tide constituents.

Hydraulics: The study of motion within large bodies of water, such as currents and waves. **Hydrology** is the study of water as it relates to the earth, such as the filtering of ground water through soil.

Interference: The interaction of simultaneous forces, such that the end result is less than the sum of the forces (at their maximal strength). For instance, when the crest of one wave occurs at the same time as the trough of a second wave, the end result is a smaller wave, or no wave at all.

Kelvin wave, coastal: A slope on the surface of moving water, caused by excess momentum of the water (acquired at a latitude closer to the equator), and the rotation of the earth beneath it. This must be distinguished from an equatorial Kelvin wave, a completely unrelated phenomenon.

Kinetic energy: The energy imparted to an object (mass) by virtue of its velocity.

Lag of the tide: See establishment.

Latitude: A line drawn around a circumference of the earth, parallel to the equator.

Linear change: Change with time that is the same during any equal interval of time, such as a car on cruise-control, that travels 60 miles every 60 minutes.

Longitude: A line drawn around the circumference of the earth, which passes through the north and south pole, and is perpendicular to the equator.

Lunar signature: A changing pattern of tides that is so obviously linked to a major lunar cycle that it allows the observer to recognize a lunar influence, such as the weekly change in lunar phases.

Mean value: In the common mathematical or scientific usage, mean is the middle point between the two extremes — between the largest and the smallest numbers or data points in a series. This is different from the average of all of the numbers in the series. However, in the terminology used in reference to the tides, mean (mean low water, e.g.) is the same as the average.

MLLW: Mean lower low water. Since the low tides are not necessarily equal on a given day, one of them may be the lower low tide for that day. The average of the lower low tides on each day during the past year is the MLLW.

MLW: Mean low water. The average of all of the low tides for the past year.

Mixed tide patterns: The erratic tide pattern that arises when neither the diurnal nor the semidiurnal celestial cycles completely dominate the tide pattern. Mixed tides may have two recognizable very unequal high tides each 24 hours, or they may have only one recognizable high tide each day. These patterns are further classified as either: (1) **Mixed, semidiurnal patterns**, in which lunar, semidiurnal tidal waves predominate, or (2) **Mixed, diurnal patterns**, in which solar, diurnal tidal waves predominate.

Modulate: To change, adjust, adapt, or vary some physical characteristic. Modulation may increase or decrease the height, or may lengthen or shorten the frequency, for example.

Month: See anomalistic month, calendar month, sidereal month, and synodic month.

Neap tides: Lower than average tides that occur during the lunar phases of both first quarter moons and third quarter moons.

Near solar system: The sun, earth, and moon. The **solar system** includes the sun and all of its planets and all of their moons.

NOAA: National Oceanographic and Atmospheric Administration. An agency of the federal government that monitors the atmosphere and oceans. **NOS** is the National Ocean Survey, an office within NOAA that produces the tide tables.

Node: The point in an oscillating system where there is no motion, such as the axis point in the center of a lake, with waves oscillating back and forth.

Nonlinear change: Change with time that is unequal during successive equal time intervals, such as a falling object that travels 12 feet during the first second, 24 feet during the next second, 36 feet during the following second, etc.

Oscillatory wave: A wave of energy moving through standing water. The surface of the water bobs up and down, but there is no net horizontal movement of water over the bottom.

Overtide: A relatively short frequency tidal wave over the continental shelf, which has an exact multiple (factor of two) of an oceanic tidal wave frequency, arriving at the mouth of that coastal basin.

Perigee: The monthly point in the elliptical lunar orbit when the moon is closest to the earth.

Perihelion: The annual point in the earth's elliptical solar orbit when the earth is closest to the sun.

Period: The time required for one complete cycle, such as the time between successive waves.

Phase of the moon: The weekly change in the appearance of the moon in the night sky — full disc, partial disc, or crescent — full moon, quarter moon, or new moon.

Phase of the tide, or phase of a tidal constituent: The period of time between the passage of the crest of the tide (or the crest of a constituent) over a point on earth, and the moment when the moon or sun is directly over that location.

Phasic: In phase. When two simultaneous forces are synchronized — for example, when they are both at their maximal strength at the same time, and reinforce each other.

Polar axis: A line through the center of the earth, from the north pole to the south pole.

Progressive wave: A wave that involves the horizontal movement of water over the bottom, such as the breakers and the water line at the beach.

Potential energy: The energy imparted to an object by moving it to a higher point above the surface of the earth, within the earth's gravitational field.

Quarter wave oscillator: An open basin with a length that is exactly one-fourth of the wavelength of the external forcing wave energy, so that the node of the wave motion is exactly at the mouth of the basin.

Range of the tide: The most common usage refers to the difference in the height of the water at low tide on a given day compared to the following high tide on that day. This is the value used by boaters, planning a day on the water. If low tide measures 0.5 feet (above MLLW), and high tide measures 6.5 feet, the range is 6.0 feet. There is also a **diurnal range**, which is the difference between the lowest low water and the highest high water on a given day. References to different ports report the **average tidal range** or **mean tidal range**, which is the difference between the average lower low water and the average higher high water during one year of tide cycles.

Reinforcement: The interaction of two simultaneous forces, such that the end result is a force greater than either of the components. For example, when two waves crest at the same time, the end result is a wave larger than either of them.

Regression of the moon's nodes: The cyclical change in the maximal lunar declination, from 18 degrees to 28.5 degrees and back to 18 degrees, throughout the longer 18.6 year cycle of changing lunar declination.

Resonance: The response of a resting mechanical system (or electrical, or atomic system) to the cyclical or periodic external forcing of energy. This occurs when the frequency of the external driving force is mathematically attuned (an even multiple, or even fraction) to the natural frequency of the resting system.

Rogue wave: An unpredictable surface wave caused by the reinforcement of multiple waves, which simultaneously crest over the same point on the bottom.

(The) Rule of twelve: A rule of thumb used to calculate the height of the water at each minute of a cycle of tidal flood and ebb. During each 6 hour flood or ebb, the tide moves 1/12th of the tidal range during the first hour, 2/12ths of the range the second hour, 3/12ths the third hour, 3/12ths the fourth hour, 2/12ths the fifth hour, and 1/12th of the tidal range during the sixth hour.

Sea level: The sea level is not level. This is a very complicated matter, which is discussed in *Beyond the Moon* on pages 89–92.

Seiche: Commonly known as the bathtub effect, a seiche is an oscillating, back and forth, series of waves within a basin.

Semidiurnal tide pattern: A tide pattern with two high tides each 24 hours, which are approximately equal — within 25% of the same height.

Shelfwaters: The seawater over the continental shelf.

Sidereal month: The time required for the moon to orbit 360 degrees around the earth — 27.32 days. This period is then adjusted to account for the rotation of the earth during one lunar orbit, to arrive at an anomalistic month.

Slack tide: The brief period of time when the water level is not changing, both at high tide and low tide.

Solar signature: A changing pattern of the tides that is so obviously linked to the major solar cycles that it allows the observer to recognize a solar influence, such as the seasonal change in tide patterns.

Spring tide: Higher than average tides that occur during the lunar phases of both new moons and full moons.

Stand in the tide: A period of time during a tide cycle in which there is no change in the height of the water, occurring between the slack tides.

Standing wave: A waveform on the surface of the water, which does not bob up and down, and does not move over the bottom. This occurs where two waves are moving in the opposite direction, and crest at the same point over the bottom. Another form of standing wave may form over a point where fast moving water passes over a very abrupt change in depth.

Sustained forcing: The transfer of more and more energy into a coastal basin by successive tidal waves, analogous to pushing a child on a swing-set in rhythm with their established motion.

Synodic month: The time required for the moon to return to the same lunar phase, such as the time between a new moon and the next new moon — 29.53 days. This is not directly linked to the daily lunar tidal wave, although it is the important determinant of another aspect of the tides, the period between successive spring tides, and successive neap tides.

Thermal expansion: As seawater heats up, it expands, raising the sea level. This has a seasonal influence on the height of the water during the tide cycles.

Tidal pumping: In shallow estuaries, the ebb tide requires more time than the flood tide. Each successive flood tide "pumps" more and more water into the coastal basin, as water is trapped within the basin during the long slow ebb. This effect is greatest during spring tides, when the strong flood tide forces in more water into the estuary.

Tidal bore: A rare condition in which the advancing flood tide arrives as a vertical collapsing wave-front, racing up a river or into a bay.

Tidal wave: The earth's oceans are deformed into an ellipse by the gravitation of the moon and the sun. There are lunar tidal waves, and solar tidal waves. The earth revolves 360 degrees each 24 hours. This causes the highest points of the oceanic ellipse (the tidal bulges) to progress from east to west around the globe. The daily movement of these tidal bulges toward the continental coasts are called tidal waves. See Tsunami.

Tidal zone: The area at the coastal waterline which is alternately covered with water at high tide, and dry land at low tide.

Tide monitoring station: A permanent installation along the coast, part of the National Water Level Observation Network (NWLON), monitored continuously by the National Ocean Survey (NOS). 175 of these stations, along the east coast of North America, and the Great Lakes, measure the height of the water with an acoustic sensor. These measurements are the basis for the tide table for that coastal location for the following year.

Tide table: A prediction of the future height and pattern of the tides at a specific location on the coast. Tide tables may either be: (1) graphic, and depicted as a curve on a graph, or (2) a tabulation, and presented as a list of the numerical height of the water at each hour.

Topography: A detailed description of the surface features of a location, the vertical elevation of the surface of the ground, either on the land or on the seafloor.

Tractal force: The horizontal gathering of the oceans around the earth toward the lunar or solar gravitation. This is distinguished from the vertical lifting of the surface of the oceans by celestial gravitation.

Tsunami: A rare, unpredictable, potentially disastrous oceanic wave caused by an earthquake, landslide, or volcanic eruption on the seafloor, or on an island, or the shoreline of a landmass.

Vector: The directional component of a force or motion. An object moving toward the northeast has a vector of motion toward the north, and another vector of motion toward the east. A force or motion is said to be **vectorial** due to the directional component.

Velocity and **Speed:** The speed of a moving object is the change in distance per unit time (miles per hour, e.g.). Velocity includes both the speed of motion and the directional vectors of the motion.

Viscosity: The measure of a liquid's resistance to flow. Honey is more viscous than water. Cold honey is more viscous than warm honey.

Wavelength: The distance between the crests of successive waves.

Wind effect: The height of the water near the coast may be elevated by wind blowing toward the shore, or lowered by wind blowing out to sea. The amount of wind effect on the height of the water depends on: (1) the speed of the wind, (2) the duration of the wind, (3) the fetch of the wind (the amount of surface effected), (4) the depth of the water, and (5) the degree to which the water is enclosed by a landmass.

Bibliography

"If you steal from one author, it's plagiarism.
If you steal from many, it's research."   Wilson Mizner

PREFACE:

Feynman, Richard P. (1995) *Six Easy Pieces*. New York: Addison-Wesley Publishing Company.

Gleick, James (1992) *Genius*. New York: Pantheon Press.

Mehra, J. (1994) *The Beat of a Different Drum*. Oxford: Oxford University Press.

CHAPTER ONE:

Bergman, Peter G. (1992) *The Riddle of Gravitation*. New York: Dover Publications, Inc.

Burn, A. R. (1948) *Alexander the Great and the Hellenistic Empire*. New York: Macmillan Company.

Cartwright, D. E. (1999) *Tides: A Scientific History*. Cambridge: Cambridge University Press.

Cohen, B. I., et al. (2002) *The Cambridge Companion to Newton*. Cambridge: University Press.

Dales, Richard (1973) *The Scientific Achievement of the Middle Ages*. Philadelphia: University of Pennsylvania Press.

Green, Peter (1991) *Alexander of Macedonia, 356–323 B. C.* Berkley: University of California Press.

Pais, Abraham (1982) *"Subtle is the Lord", The Science and Life of Albert Einstein.* Oxford: Oxford University Press.

Sobel, Dava (1999) *Galilleo's Daughter.* New York: Walker and Company.

White, Michael (1999) *Isaac Newton, The Last Sorcerer.* Reading, Mass: Persius Publishing.

Wylie, F. E. (1979) *Tides and the Pull of the Moon.* Battleboro, Vermont: The Stephen Greene Press.

CHAPTER TWO:

Hansen, W. (1962) Tides. In: Hill, M. N. (ed) *The Sea, Ideas and Observations in the Progress in the Study of the Seas.* pp. 764–801. New York: Interscience Publisher.

Pugh, D. T. (2001) Tides. In: Steele, John H. (ed) *Encyclopedia of Ocean Sciences.* Vol 6. 2961–2968.

Schoettle, T. (1906) *A Guide to a Georgia Barrier Island.* St. Simons Island, Georgia: Watermarks Publishing.

Smith, Raymond, et al. (1926) *Astronomy: A Revision of Young's Manual of Astronomy.* Boston: Ginn and Company.

Swartz, Clifford E. (1998) *Teaching Introductory Physics.* New York: Springer-Verlag, pp. 101–133.

Wells, N. (1986) *The Atmosphere and Ocean.* London: Taylor and Francis.

CHAPTER THREE:

Goldreich, P. (1972) Tides and the Earth–Moon System. *Scientific American.* April: 42–57.

NOAA, *Our Restless Tides*:
http://coops.nos.noaa.gov/restless4.html.

McLellen, H. J. (1965) *Elements of Physical Oceanography.* Oxford: Pergammon Press.

Smith, Raymond, et al. (1926) *Astronomy: A Revision of Young's Manual of Astronomy.* Boston: Ginn and Company.

Wylie, F. E. (1979) *Tides and the Pull of the Moon.* Battleboro, Vermont: The Stephen Greene Press.

CHAPTER FOUR:

Darwin, G. H. (1962) *The Tides and Kindred Phenomena of the Solar System.* San Francisco: W. H. Freeman.

Defant, Albert (1961) *Physical Oceanography.* Vol. 2. pp. 245–352. Oxtord: Pergammon Press.

Duxbury, A. C. (1971) *The Earth and Its Oceans.* pp. 317–340. Reading, Massachusetts: Addison-Wesley.

Swartz, Clifford E. (1998) *Teaching Introductory Physics.* New York: Springer- Verlag. pp. 101–133.

CHAPTER FIVE:

Culver, C. A. (1956) *Musical Acoustics.* New York: McGraw-Hill.

Mulligan, J. F. (1985) *Introduction to College Physics.* pp. 301–302. New York: McGraw-Hill.

Pond, S., and Pickard, G. L. (1978) *Introductory Dynamic Oceanography.* Oxford: Pergammon Press.

CHAPTER SIX:

Amphidromic systems... co-oscillations in the English channel... the damping of the reflected Kelvin wave... the tide moves as a... www.bangor.ac.uk/~oss110/oxp2003/lecture4/pdf.

Becker, R. A. (1954) *Introduction to Theoretical Mechanics.* N. Y.: McGraw-Hill. pp. 248–253.

Cuchlaine, A. M. (1963) *An Introduction to Oceanography.* New York: McGraw-Hill.

Defant, Albert (1961) *Physical Oceanography.* Vol. 2. pp. 320–322. Oxford: Pergammon Press.

Duxbury, A. C. (1971) *The Earth and Its Oceans.* pp. 316–340. Reading, Massachusetts: Addison-Wesley.

Feynman, Richard P. (1966) *The Feynman Lectures on Physics.* Reading Mass.: Addison-Wesley. Sections 18-1 to 19-9.

Goldstein, Herbert. (1980) *Classical Mechanics.* Reading Mass.: Addison-Wesley. pp.161–184.

Gorlov, A. M. (2001) Tidal Energy. In: Steele, John H. (ed) *Encyclopedia of Ocean Sciences.* Vol. 6. pp. 2955–2961.

Henderson and Speranza (1971) *Deep Sea Resonance,* 18: 959–980.

MacDonald, J. E. (1952) The Coriolis Effect. *Readings from Scientific American — Oceanography.* San Francisco: W. H. Freeman Company.

Marion, Jerry B. (1988) *Classical Dynamics.* New York: Harcourt Brace. pp. 23–35, pp. 43–50, pp. 336–345.

Resnick, Robert. (1979) *Physics, ed 4.* New York: John Wiley and Sons. pp.118–120.

Sommerfield, Arnold (1964) *Mechanics.* London: Academic Press. pp.162–174.

Stommer, Henry M. (1989) *Introduction to the Coriolis Force.* New York: Columbia University Press.

Swartz, Clifford E. (1998) *Teaching Introductory Physics.* New York: Springer-Verlag. pp. 101–133.

Synge, John L. (1949) *Principles of Mechanics.* New York: McGraw-Hill. pp. 148–157, pp. 349–351.

Taylor, (1920) *Procedures of the London Mathematical Society*, Vol. 20, pp. 144–181.

Wells, N. (1986) *The Atmosphere and Ocean.* London: Taylor and Francis.

CHAPTER SEVEN:

Bowden, K. F. (1983) *Physical Oceanography of Coastal Waters.* New York: Halsted Press (John Wiley and Sons).

Chapman, D. C. and Giese, G. S. (2001) Seiches. In: Steele, John H. (ed) *Encyclopedia of Ocean Sciences.* Vol. 5. pp. 2724–2731. San Diego: Acedemic Press.

Defant, Albert (1961) *Physical Oceanography.* Vol. 2. p. 338. Oxford: Pergammon Press.

Korgen, B. A. (1996) Seiches. *American Scientist.* 83: 330–341.

Zetler, B. D. (1972) Tides in the Gulf of Mexico. In: Capurro, L. R. A. (ed) *Contributions to the Physical Oceanography of the Gulf of Mexico.*l Vol. 2.2.Houston, Texas: Gulf Publishing Company.
www.physics.ohiostate.edu/~dvandom/Edu/newcor.html.

CHAPTER EIGHT:

Bartholomew, W. T. (1942) *Acoustics of Music.* New York: Prentice-Hall.

Chapman, D. C. (1996) A Model for the Generation of Coastal Seiches by Deep-Sea Internal Waves. *Journal of Physical Oceanography.* 20: 1459–1467.

Duxbury, A. C. (1971) *The Earth and Its Oceans.* Reading, Massachusetts: Addison-Wesley.

Holman, R. A. (2001) Waves on Beaches. In: Steele, John H. (ed) *Encyclopedia of Ocean Sciences.* Vol. 6. pp. 3194–3201.

Honda, H. (1996) Secondary Undulations of Oceanic Tides. *Journal of the College of Science, Imperial University,Tokyo.* 24: 1–113.

Kowalik, Z. (1995) Tides in the Sea of Okhotsk, *Journal of Physical Oceanography*, 28: 1389–1409.

Miles, A. B. (1974) Harbor Seiching. *Annual Review of Fluid Mechanics.* 6: 17–35.

Redfield (1958) *Journal of Marine Research*, 17: 432–448.

Simpson, J. H. (1998) Tidal Processes in Shelf Seas. In: Brink, K. H. (ed) *The Sea.* Vol. 10. New York: John Wiley.

Wells, N. (1986) The Atmosphere and Ocean. London: Taylor and Francis.

Winchester, Simon. (2004) *Krakatoa.* New York: Harper Collins.

Wylie, F. E. (1979) *The Tides and the Pull of the Moon.* Battlesboro, Vermont: The Stephen Greene Press.

CHAPTER NINE:

Bowden, K. F. (1983) *Physical Oceanography of Coastal Waters.* New York: Halsted Press (John Wiley and Sons).

Officer, C. B. (1976) *Physical Oceanography of Estuaries.* New York: John Wiley and Sons.

Wickstrom, Carl. et al. (2004) *Florida Sportsman Fishing Planner.* Stuart, Florida. The Florida Sportsman.

Wylie, F. E. (1979) *The Tides and the Pull of the Moon.* Battleboro, Vemont: The Stephen Greene Press.

Estuaries of New South Wales. Physical Characteristics and Behavior:
http://www.dlwc.nsw.gov.au/care/water/estuaries/FactSheets/Physical/tidal-behavior.html.

Personal communication: Dr. Keven Bodge, Olsen and Associates, Inc., Jacksonville, Florida.

CHAPTER TEN:

Defant, Albert (1961) *Physical Oceanography.* Oxford: Pergammon Press.

Gross, M. (1971) *Oceanography.* pp. 111–121. Columbus, Ohio: Charles Merrill Publishing Company.

Knauss, J. A. (1978) *Introduction to Physical Oceanography.* pp. 229–241. Englewood Cliffs, N. J.: Prentice-Hall, Inc.

NOAA, Tide Predicting Machines:
http://www.co-ops.nos.noaa.gov/tide predicting machines. html.

Our Restless Tides:
http://www.co-ops.nos.noaa.gov/restles 1.html.

Parker, B. B. (ed) (1991) *Tidal Hydrodynamics.* New York: John Wiley.

Parker, Sybil., et al. (1980) *The McGraw-Hill Encyclopedia of Ocean and Atmospheric Science.* New York. McGraw-Hill.

Ray, R. D. and Woodworth, P. L. (ed) (1997) Special Issue on Tidal Science in Honor of David Cartwright. *'Progress in Oceanography.* 40.

Wells, N. (1986) *The Atmosphere and Ocean.* London: Taylor and Francis.

Personal communication: Mr. Stephen Gill, office of the NOS, NOAA.

CHAPTER ELEVEN:

Folland, C. K. (1990) *Observed Climate Variations and Change: The IPPC Scientific Assessment.* ed. By Houghton, J. T. et al. pp. 195–238. Cambridge.

McBean, G. A. (1990) *Global Energy and Water Fluxes.* Weather, Vol. 44. pp. 255–291.

Schneider, S. H. (1996) *Encyclopedia of Climate and Weather.* New York: Oxford Press. pp. 83–84, 761–763.

Wells, N. (1986) *The Atmosphere and Ocean.* London: Taylor and Francis.

Oscillation in barometric pressure, resulting from the passage of weatherfronts.
www.netl.doe.gov/products/em/indUnivProg/pdf/2307.pdf.

U. S, National Report to IUGG, 1991–1994, *Rev. Geophysical Union, Vol. 3 Supplement,* American Geophysical Union, 1995.

CHAPTER TWELVE:

Carson, Rachel. *The Edge of the Sea.* New York: Houghton Mifflin, 1998.

Earnhardt, Thomas W., *Flyfishing the Tidewaters.* New York: Lyons and Burford, Publishers, 1995.

Roberts, George V., A Fly-fisher's Guide to Saltwater Naturals and Their Imitation. Camden, Maine, 1994.

Tabory, Lou., *Inshore Fly Fishing.* New York: Lyons and Burford, Publishers, 1992.

CHAPTER THIRTEEN:

Brunsden, D., et al. (1999) <u>Atlas of the World</u>. Oxford: Oxford University Press.

Gleick, James (1996) *Chaos: Making a New Science.* New York: Pantheon Press.

Melchior, Paul, (1978) *The Tides of the Planet Earth*: New York: Pergammon Press.

EPILOGUE:

Gulf of Maine: http://www.gomoss.org/images/aboutgom/map.

Steele, John H. (2001) Appendix 10, Bathymetric Charts of the Oceans.

The Encyclopedia of Ocean Science. Vol. 6. pp. 3296–3303. San Diego: Academic Press.

Index

Age of the tide, 89
Alexander the Great, 1
Amazon River, 155
Amphidromes, 72, 73, 79, 81–84, 91, 167
Amplification by resonance, 129–133
Amplitude of wave, 119, 120
Anomalies from tide table, 170, 175
Anomalistic month, 42, 43, 61
Aphelion of the earth's orbit of the sun, 48, 49, 245
Apogee of the lunar orbit, 38, 49, 245
Astronomical influences on tides, 22, 24, 27, 33, 38, 39, 43, 47, 51, 57, 61
Atlantic ocean, 91, 108, 109, 233, 253
Atmospheric pressure, 185–188
Atmospheric tides, 191–193

Barometric pressure, 77, 97, 134, 185–188
Basins of oscillation, 97, 98, 108, 110, 247
Bathymetric chart, 244, 256
Bay of Fundy, 123, 137–140
Bermuda, 110
Bernoulli, Daniel, 5
Biology and tides, 198–200
Bohr, Neils, 145
Bore, tidal, 154–156
Bristol Channel, England, 122, 140

Calendar, 40
Cape Hattaras, 124
Caribbean Islands, 82
Centrifugal force, 20, 23–25, 27, 29, 93, 94

Centrifugal force, earth deformity due to, 29
Chaos theory, 173–181
Chesapeake Bay, 171
Climate change, 176, 238
Coastline, contour of, 122–124
Complexity, 65, 173, 176, 181
Conservation of angular momentum, 94
Constituents of the tide, 160, 217, 226
Continental shelf, 105–107, 126
Cooks Inlet, Alaska, 137, 140, 256
Co-oscillating tides, 142, 143
Co-range lines, 81, 83
Coriolis effect on ocean currents, 87
Coriolis effect on tidal currents, 85, 86
Coriolis force, 74–80, 93, 94
Coriolis force and atmosphere, 78
Coriolis force and sea level, 80
Coriolis, Gaspard Gustave de, 74
Co-tidal lines, 73, 81, 83
Current, ocean, 87, 91
Current, tidal, 85–87
Currents, tidal, 200, 202

Declination of equator to solar orbit, 51, 52, 67
Declination of the lunar orbit, 33, 34, 41, 65, 67
Diurnal tides, 51, 61, 67, 104
Doodson, Arthur, 7, 11
Dynamic Theory of the Tides, 6, 97

Earth orbit of lunar orbital axis, 19, 22, 23
Earth orbit of sun, 49, 50
Earth, rotation of, 24, 75, 159

Easter, 40
Ebb tide, 15, 37, 38, 147, 148
Einstein, Albert, 12, 31, 45, 114
Electrical generator, tidal, 88
Elliptical deformity of oceans, 19, 23, 25
English Channel, 84, 140
Equator, 32, 33, 51
Equilibrium Theory of the Tides, 5, 57
Establishment of the tide, 89, 161, 162
Estuary, 146–151, 153

Fetch of the wind, 190
Feynman, Richard, 19, 71
Fishing and barometric pressure, 188–190
Fishing and currents, 200, 202
Fishing and lunar phases, 208
Fishing and tidal range, 201
Fishing and water temperature, 211–213
Fleisher–Harris tide machine, 157
Flood tide, 15, 38, 147, 148
Florida, 124, 249
Fortnightly effect, 53, 61
Fortnightly tide pattern, 150
Fourier analysis, 169
Frame of reference, 25, 73, 81
Frequency, of tidal wave, 100
Friction, 118, 146
Fu Ch'un River, China, 155
Full moons, 47

Galileo Galilei, 4, 31
Galveston Bay, 171
Geographic influences on tides, 122–125, 131, 162
Geoid, 90
Geostrophic forces, 57, 72, 159, 162, 226
Gravitation, 7, 12, 21
Gravitation, earth deformity due to, 56
Gravitation, lunar, 10, 28, 35, 38
Gravitation, solar, 10, 48, 50

Gulf of Maine, 123, 251
Gulf of Mexico, 92, 103
Gulf of Thailand, 101

Half tide level, 149
Half-wave oscillator, 134
Harmonic analysis, 165–170
Harmonic theory, 7, 226
Hawaiian Islands, 83
Height of tide, 223, 224
High tide, 66, 89
Highest tides of year, 52
Highest tides on earth, 35, 139, 140
Hydraulic influences on tides, 84, 106, 116–121, 129–137, 162

Independent tides, 142
Indian ocean, 140
Interference, 42, 62, 64, 65, 68, 104, 171
Islands tides, 110

K1, tide constituent, 110, 168, 217
Kelvin waves, coastal, 84, 85
Kelvin, William Thompson, Lord, 5
Kepler, Johannes, 4, 60
Kimberley Coast, Australia, 140
Kinetic energy, 37, 119
King's Sound, Australia, 126
Krakatoa, 120

La Dordogne River, France, 156
Lag of the tide, 89, 161
Lagrange's formula, 137
Lake Geneva, 97
Laplace, Pierre-Simon, 5, 97
Light waves, 102
Low pressure centers, atmospheric, 77, 78
Low tide, 66, 90
Lunar gravitation, 27, 28, 35, 37, 38, 48
Lunar nodal tide, 34
Lunar orbit, 22, 32, 33, 39, 40, 50
Lunar phases, 46, 47

Lunar signatures, 53
Lunar tide, 150, 237

M2, tide constituent, 168, 217
Maelstrom, 40
Mayport, Florida, 124
Mean low water (MLW), 90
Mean lower low water (MLLW), 90
Mean sea level (MSL), 90
Mediterranean, 2, 3, 142
Merian's formula, 137
Miami, Florida, 124
Middle ages, 4
Minas Basin, Nova Scotia, 123, 155
Mixed tides, 66, 67, 109, 233
Month, anomalistic, 42
Month, calendar, 42
Month, sidereal, 42
Month, synodic, 43

National Geodetic Vertical Datum (NGVD), 91
National Ocean Survey (NOS), 164
National Oceanographic and Atmospheric Administration (NOAA), 7, 11, 157, 158, 164
National Tide Datum Convention, 90
National Tide Datum Epoch, 90
National Water Level Observation Network (NWLON), 164, 165
Neap tide, 46, 48, 149, 150, 236, 237
Near solar system, 49
New moons, 47
Newton, Isaac, 4, 5, 31, 45
Newton's Laws of Motion, 93
Node of oscillation, 133, 134

Ocean currents, 78, 87, 91
Oceanography, influences on tides, 72, 80, 108
Okhotsk Sea, Russia, 140
Orbital systems, 19–21, 49
Oscillatory waves, 85, 107, 117
Overtides, 138

Pacific ocean, 91, 108, 140, 233, 253
Pattern of tides, 35, 225
Perigee of the lunar orbit, 38, 49, 245
Perihelion of the earth's orbit of the sun, 49, 245
Period of oscillation, 132, 137
Period of tide, 15, 135
Period of wave, 121
Phase of tide, 89, 161
Phases, lunar, 46, 47
Polar tides, 34
Potential energy, 37, 119
Precession of the earth, 176
Prince William Sound, 10
Progressive wave, 107, 117
Proudman, Joseph, 7
Pythagoras, 59

Quarter moons, 47
Quarter wave oscillator, 134

Rance estuary, 85, 88
Range of tide, 36, 53, 201
Refraction of wave, 118
Regression of the moon's nodes, 34, 61
Reinforcement, 42, 62, 64, 65, 68, 171
Resonance, 129–133
Response Analysis Theory of the Tides, 6
River, tides in, 87, 153, 154
Rogue waves, 99, 100
Rotation of earth, slowing of, 88
Rotation of the earth, 24, 39, 72, 75, 79, 86, 88, 159, 176
Rule of Twelve, 201, 205

S2, tide constituent, 217
Sea level, 89–92
Seasonal tides, 52
Sedimentation, 146, 148
Seiche effect, 97, 98, 104–106, 108
Semidiurnal tides, 24, 33, 51, 61, 66, 67, 109, 233

Severn River, 85
Shelfwaters, 105, 106
Sidereal month, 42
Solar declination, 51–53, 67, 244
Solar gravitation, 36, 48–50
Solar orbit, 49, 50
Solar radiation and tides, 52
Solar signature, 53
Solar system, 17, 49
Solar tide, 47, 150, 237
Solunar tables, 214
Sound waves, 8, 59, 116, 129
Spring tide, 46, 47, 61, 149
Stand in the tide, 69, 105, 151
Standing waves, 105–107
Surge, 118, 119
Sustained forcing, 135–137, 143
Synodic month, 43

Tahiti, 110
Tectonic plates, 176, 238
Temperature, 52, 146
Thermal expansion, 52
Thermotidal oscillation, 192
Tidal bore, 154–156
Tidal current, 85–87, 200, 202
Tidal electrical generator, 88
Tidal intermixing, 98, 103, 104, 106, 133
Tidal phase, 89
Tidal pumping, 149, 150
Tidal range, 201
Tidal remixing, 136
Tidal trapping, 150, 153
Tidal wave, 62
Tidal wave, constituents of, 99–103
Tidal wave, frequency of, 100, 102, 161
Tidal wave, phase, 161
Tidal wave, reflection of, 98, 104, 106, 133
Tidal wave, speed of, 37
Tidal wave, time of arrival, 82, 129
Tidal wave, wavelength of, 108
Tidal winds, atmospheric, 192

Tide corrections, 151, 152
Tide monitoring station, 165
Tide patterns, 35, 66, 108, 235
Tide powered electric generators, 88
Tide predicting machines, 157, 158
Tide table anomalies, 170, 175
Tide table computation, 160–172
Tide, astronomical influences on, 22, 27, 38, 39, 43, 47, 51, 61
Tide, constituents of, 99, 160, 217, 226
Tide, daily shift of, 40
Tide, diurnal, 51, 66, 67
Tide, ebb, 15, 38, 147, 148
Tide, establishment of, 89
Tide, flood, 15, 38, 147, 148
Tide, geographic influences on, 122–125, 162
Tide, height of, 223, 224
Tide, high, 66, 89
Tide, hydraulic influences on, 84, 116–121, 162
Tide, low, 66
Tide, lunar, 150, 237
Tide, neap, 46, 149
Tide, oceanography influence, 72, 80, 82, 108
Tide, offshore, 150
Tide, pattern of, 35, 225
Tide, period of, 15, 135
Tide, semidiurnal, 24, 51, 61, 66, 67, 109, 233
Tide, solar, 47, 150, 237
Tide, spring, 46, 47, 149
Tides and biology, 198–200
Tides and fishing, 197
Tides on islands, 110
Tides, atmospheric, 191–193
Tides, highest of year, 52
Tides, highest on earth, 35, 139, 140
Tides, mixed, 67, 109, 233
Tides, polar, 34
Tides, seasonal, 52
Topography, 122, 125, 126, 256
Tractal forces, 55
Tsunami, 62, 120, 128

Viscosity, 146, 147

Wave, amplitude, 118–120
Wave, deep water, 37
Wave, oscillatory, 85, 107, 117
Wave, period, 121
Wave, progressive, 107, 117
Wave, refraction, 118
Wave, speed of, 137

Wavelength, 120
Waves, light, 102
Waves, sound, 8, 59, 116, 129
Weather, 161, 171, 185, 239
Weather and tides, 185–193, 226
Whirlpool, 40, 78, 81
Wind and water level, 190–193
Wylie, F. E., 7